DIOPHANTINE EQUATIONS AND SYSTEMS: An Introduction

A systematic approach to problem solving

The Pythagorean Equation

$$x^2 + y^2 = z^2$$

$$\{x = (m^2 - n^2)k, \quad y = 2mnk \quad z = (m^2 + n^2)k\}$$

$$(m, n, k \in \mathbb{Z})$$

The book contains 55 solved illustrative exercises and 105 problems for solution. Odd numbered problems are provided with answers. Hints or detailed outlines are given for the more involved problems.

Diophantine Equations and systems: An Introduction

Demetrios P. Kanoussis, Ph.D

PREFACE

Diophantine equations are polynomial equations with **integer coefficients** for which **only integer solutions** are sought. In his great work "**Arithmetica**", the Greek mathematician **Diophantus of Alexandria**, (born in Alexandria Egypt in 200 AD and died in 284 AD), known as the father of Algebra, studied and solved such types of equations, (integer coefficients and integer solutions), of the first up to the fourth degree. These equations are now known as "**Diophantine equations**". A characteristic feature of Diophantine equations is that in these equations the number of equations is smaller than the number of unknowns. For example, we may have one equation with two unknowns, or one equation with three unknowns, or a system of two equations with three unknowns, etc. While in the set of real numbers $\mathbb{R}$ these types of equations, (fewer equations than number of unknowns), are indeterminate, in the set of integers $\mathbb{Z} = \{\dots -3, \ -2, -1, 0, 1, 2, 3, \dots\}$ or in the set of natural numbers $\mathbb{N} = \{1, 2, 3, 4, \dots\}$, these equations may or may not have integer solutions, (depending on the coefficients of the equations).

In this book we provide a systematic introduction to Diophantine equations, with emphasis on the solution of various problems. The first two chapters are devoted to first degree Diophantine equations and systems, (linear equations and systems), while the third chapter is devoted to second degree Diophantine equations and systems. Among other equations, in this chapter, we study the Pythagorean equation $(x^2 + y^2 = z^2)$, and the Pell's equation $(x^2 - ky^2 = 1)$. The solution of Pell's equation is achieved by a really brilliant method, which is attributed to **Lagrange**. Various examples of higher degree Diophantine equations are considered in chapter 4. The analytic description of the material covered in this book can be found in the table of contents.

The book is concluded with a collection of 40 miscellaneous, challenging problems, with answers and detailed remarks and hints.

In total, the book contains 55 solved examples and 105 problems for solution.

TABLE OF CONTENTS

CHAPTER 1: FIRST DEGREE DIOPHANTINE EQUATIONS

CHAPTER 2: DIOPHANTINE SYSTEMS OF LINEAR EQUATIONS

CHAPTER 3: SECOND DEGREE DIOPHANTINE EQUATIONS

CHAPTER 4: HIGHER DEGREE DIOPHANTINE EQUATIONS (EXAMPLES)   74

CHAPTER 1: FIRST DEGREE DIOPHANTINE EQUATIONS

1-1) Introduction

A Diophantine equation is a polynomial equation that contains two or more unknowns, in which all the coefficients are integers and all the solutions of interest are integers as well.

In other words, the equation $f(x, y, z, \ldots) = 0$ is said to be "**a Diophantine equation**", if $f(x, y, z, \ldots)$ is a polynomial in $x, y, z, \ldots$, **having all its coefficients integers and if we seek only for integer solutions** , i.e. integer numbers $a, b, c, \ldots$, such that when $x, y, z, \ldots$ are replaced by the integers $a, b, c, \ldots$, respectively, the equation is satisfied, i.e. $f(a, b, c, \ldots) = 0$.

The degree of the polynomial $f(x, y, z, \ldots)$ is the degree of the Diophantine equation.

A Diophantine equation has fewer equations than unknowns.

The study of Diophantine equations is termed **Diophantine analysis**.

Examples: 1) The equation $2x + 3y = 10$ is an equation **of first degree, with two unknowns, x and y**. Solving this Diophantine equation means to find **all the integer solutions**, i.e., all the integer numbers $x = x_0$ and $y = y_0$, (with x_0 and y_0 integers), which satisfy the equation. **A Diophantine equation may or may not have integer solutions**.

In general, the equation $\boldsymbol{ax + by = c}$, where a, b and c are integers, is called "**a linear Diophantine equation of first degree , in two unknowns, x and y**". We shall study this equation in the next section. As we shall see, this equation may or may not have integer solutions, (depending on the integer coefficients a, b and c).

2) The equation $\sqrt{2}\, x + 5y = 9$, is **not** a Diophantine equation, since the coefficient of x is not integer, ($\sqrt{2}$ is an irrational number).

3) The equation $7\sqrt[3]{x} - 2y = 15$, is **not** a Diophantine equation, since x is not raised to a positive exponent, i.e., $7\sqrt[3]{x} - 2y$ is not a polynomial.

4) The equation $x^2 + y^2 = z^2$ is **a second degree Diophantine equation**. Its solutions are known as "**the Pythagorean triples**". For example, the numbers $(x = 3, y = 4, z = 5)$ is a Pythagorean triple, since $3^2 + 4^2 = 5^2$. This is not the only Pythagorean triple. In section 3-2 (b) we shall show that **the Pythagorean equation has an infinite number of integer solutions**, i.e. that there are infinitely many Pythagorean triples, and we shall find their general expression.

5) The equation $2x + 6y + 10z = 18$ is **a linear Diophantine equation in three unknowns**.

6) The system of the two equations

$$\begin{cases} 5x + 7y - 3z = 10 \\ 2x - 9y + 4z = 18 \end{cases}$$

is **a Diophantine system of two, linear equations, in three unknowns**.

1-2) Diophantine equations of the first degree in two unknowns, $(ax + by = c)$

a) Let us consider the Diophantine equation $ax + by = c$, where a, b and c are integer numbers, $(a, b, c \in \mathbb{Z})$, and $a \neq 0, b \neq 0$.

Solution of the Diophantine equation is any pair of integer numbers (x_0, y_0) that satisfies the equation, (i.e. $ax_0 + by_0 = c$).

Solving the Diophantine equation means to find **all the integer solutions** of the equation.

For example, the pair of integers $(2,1)$ is a solution of the equation $3x + 7y = 13$, since $3 \cdot 2 + 7 \cdot 1 = 13$. Another solution of the same equation is the pair $(9, -2)$, since $3 \cdot 9 + 7 \cdot (-2) = 13$.

On the other hand, the equation $2x + 4y = 7$ does not have any integer solutions.

In general, as we shall show in the sequel, **the Diophantine equation** $ax + by = c$, **either has an infinite number of integer solutions, or, it does not have any integer solutions at all.**

b) The following theorems are important for the solution of the Diophantine equation $ax + by = c$.

Theorem 1-1: If a and b have a common divisor q, which does not divide c, then the equation $ax + by = c$ does not have any integer solutions.

Proof: The integer q divides both a and b, (by assumption). If the equation admits an integer solution (x, y), then, q must divide ax, (since it divides a) and also q must divide by, (since it divides b), and therefore q must divide the sum $ax + by$, i.e. q must divide the number $c = ax + by$. But this contradicts our hypothesis, that q does **not** divide c, and this of course means that the equation is not satisfied by integers x and y, i.e. the Diophantine equation does not have integer solutions.

Theorem 1-2: Let d be the Greatest Common Divisor of a and b, i.e. let $d = (a, b)$. Then, the Diophantine equation $ax + by = c$ has an integer solution, if and only if, d divides c.

Proof: Since d divides a, b and c, if we divide both sides of the equation by d, we find the equivalent equation

$$Ax + By = C, \qquad where \;\; A = \frac{a}{d}, B = \frac{b}{d}, C = \frac{c}{d} \qquad (*)$$

Notice that A, B and C are **integers**, and that furthermore, A and B are relatively prime integers, i.e. $(A, B) = 1$, (recall that, when two integers are divided by their G.C.D, the integers become relatively prime). We may also assume that $A > 0$, since if $A < 0$, we may multiply both sides of the equation by (-1) and obtain the equivalent equation: $(-A)x - By = -C$, with $-A > 0$.

Solving eq. (*) for x we find:

$$x = \frac{C - By}{A} \qquad (**)$$

Let us now assume that y takes on, successively, the integer values $\{0, 1, 2, \ldots, A - 1\}$. We shall show that for **one** of these values of y, the corresponding number $(C - By)/A$ is an integer. So, for this particular, integer value of y, the corresponding value of x will be integer as well. In other words, we have found a pair of integers (x, y) which satisfies eq. (**), and therefore eq. (*), and this means that the Diophantine equation $ax + by = c$ has an integer solution.

We start the proof by showing first, that the remainders of the divisions $(C - By)/A$ are distinct, i.e. **different from each other**, when y takes on, successively, values from the set of integers, $\{0, 1, 2, \ldots, A - 1\}$.

Indeed, if two values of y from the set $\{0, 1, 2, \ldots, A - 1\}$, say the values $y = m$ and $y = n$, rendered the same remainder, say r, then we would have, (according to the equality of the algorithmic division):

$$\left.\begin{array}{c} C - Bm = q_1 A + r \\ and \\ C - Bn = q_2 A + r \end{array}\right\} \quad where \quad 0 \le r < A, \quad and \quad q_1, q_2 \in \mathbb{Z} \quad (***)$$

Subtracting the second from the first equation yields:

$$(n - m)B = (q_1 - q_2)A \qquad\qquad (****)$$

Since A divides the number $(n - m)B$, and is relatively prime to B, it must divide $(n - m)$. But, since $0 \le n < A$ and $0 \le m < A$, it follows that $-A < n - m < A$, i.e. $|n - m| < A$, and therefore A cannot divide the number $(n - m)$ which is smaller than A. We are therefore forced to accept that all the remainders of the divisions $(C - By)/A$, when y takes on values from the set $\{0, 1, 2, \ldots, A - 1\}$ are **all different from each other**. The possible remainders of the divisions $(C - By)/A$ are $\{0, 1, 2, \ldots, A - 1\}$, and since they are all different from each other, **one of them must necessarily be zero**. Let us therefore assume that for a value y_0 of the set $\{0, 1, 2, \ldots, A - 1\}$ the remainder of the division $(C - By_0)/A$ is zero. This means that the corresponding value of x, say x_0, from eq. (**) is integer as well. We have thus shown that the given equation admits **a pair (x_0, y_0) of integer solutions**, and this completes the proof.

Next theorem shows that if $ax + by = c$ has one integer solution, then it has an infinite number of integer solutions.

Theorem 1-3: If the Diophantine equation $ax + by = c$ has one integer solution (x_0, y_0), then it has an infinite number of integer solutions, given by the formulas:

$$\left\{ x = x_0 + \frac{b}{(a,b)} k, \quad y = y_0 - \frac{a}{(a,b)} k \right\}, where\ k = 0, \pm 1, \pm 2, \pm 3, ...$$

$$(1 - 2 - 1)$$

Poof: Assume that (x_0, y_0) is a solution of the Diophantine equation $ax + by = c$. This means that $ax_0 + by_0 = c$. We check easily that x and y as given in eq. (1-2-1), satisfy the equation. Indeed:

$$a \left(x_0 + \frac{b}{(a,b)} k \right) + b \left(y_0 - \frac{a}{(a,b)} k \right) =$$

$$ax_0 + \frac{ab}{(a,b)} k + by_0 - \frac{ab}{(a,b)} k = ax_0 + by_0 = c$$

Comments:

1) If a and b are relatively prime, i.e. if $(a, b) = 1$, the Diophantine equation $ax + by = c$ has always integer solutions, since $1 = (a, b)$ always divides c, (Th. 1-2).

2) If a and b are relatively prime, i.e. if $(a, b) = 1$, the formulas in equation (1-2-1) are simplified to the following:

$$x = x_0 + bk, \quad y = y_0 - ak, \quad k = 0, \pm 1, \pm 2, \pm 3, ... \quad (1 - 2 - 2)$$

3) If a and b are not relatively prime, then we may use the formulas in equation (1-2-1), or, alternatively, we may divide both sides of the equation by their G.C.D (a, b), and then apply the formulas in eq. (1-2-2). **In both cases, a first solution (x_0, y_0) must be known.**

4) If we seek for **the positive integer solutions** of a Diophantine equation, with $(a, b) = 1$, then the values of k which yield positive solutions are found from **the simultaneous solution** of the inequalities:

$$\begin{Bmatrix} x > 0 \\ and \\ y > 0 \end{Bmatrix}, \quad or, \quad \begin{Bmatrix} x_0 + bk > 0 \\ and \\ y_0 - ak \end{Bmatrix} \qquad (1-2-3)$$

The method of solution is illustrated in the following examples.

Example 1-2-1: Solve the Diophantine equation: $2x + 3y = 7$.

Solution: Since 2 and 3 are relatively prime, the equation has integer solutions, (see comment 1). Solving for x we find:

$$x = \frac{7 - 3y}{2} \qquad (*)$$

We assign y the values $0, 1$. According to Theorem 1-2, for one of these values, the corresponding x will be integer.

For $y = 0$, $x = 7/2$, (not integer). For $y = 1$, $x = (7 - 3)/2 = 4/2 = 2$, (integer). Thus, one solution of the equation is ($x_0 = 2, y_0 = 1$). Indeed, notice that $2 \cdot 2 + 3 \cdot 1 = 7$. The **general solution** of the equation is given by the formulas in eq. (1-2-2):

$$x = 2 + 3k, \quad y = 1 - 2k, \quad k = 0, \pm1, \pm2, \pm3, \ldots \qquad (**)$$

For example, for $k = 1$, we find the solution ($x = 5, y = -1$). For $k = -2$, we find another solution, ($x = -4, y = 5$), etc.

Example 1-2-2: In example 1-2-1 find the positive solutions.

Solution: We want $x > 0$ and $y > 0$, i.e.

$$\begin{Bmatrix} 2 + 3k > 0 \\ and \\ 1 - 2k > 0 \end{Bmatrix}, \quad i.e. \quad \begin{Bmatrix} k > -2/3 \\ and \\ k < 1/2 \end{Bmatrix}$$

The only integer k, between $-2/3$ and $1/2$ is $\boldsymbol{k = 0}$, and thus, the only positive solution, (from eq. (**) in example 1-2-1), is: $x = 2, y = 1$.

Example 1-2-3: Find all the integer solutions of the equation $14x + 20y = 30$. Does this equation have positive solutions?

Solution: The G.C.D of 14 and 20 is 2, and since 2 divides 30, the equation admits integer solutions, (Th. 1-2). Dividing both sides by 2 we obtain an equivalent equation, (i.e. one having the same solutions with the original), but with smaller coefficients. We find, $7x + 10y = 15$, or

$$x = \frac{15 - 10y}{7} \qquad (*)$$

We try $y = 0,1,2,3,4,5,6$. One of these values will yield an integer x, (see Th. 1-2). We find that when $y = 5$ the corresponding $x = -5$. All other values of y will make the corresponding x, in eq. (*), to be a fraction and not an integer. So, we have found one integer solution, $(x_0 = -5, y_0 = 5)$. The general solution is given by the formulas in eq. (1-2-2):

$$x = -5 + 10k, \quad y = 5 - 7k, \quad k = 0, \pm 1, \pm 2, \pm 3, \dots \dots \qquad (**)$$

Regarding the positive solutions, we must have, $x > 0$ **and** $y > 0$, i.e. $-5 + 10k > 0$ **and** $5 - 7k > 0$. The first inequality is satisfied when $k > 5/10$, i.e. $k > 1/2$, while the second inequality is satisfied when $k < 5/7$, and since there are no integer values of k between $1/2$ and $5/7$, we conclude that the equation does not have positive, integer solutions.

Example 1-2-4: Solve the Diophantine equation $117x + 278y = 1068$.

Solution: a) Since 117 is odd and 278 is even, their G.C.D is 1, and this means that the equation does have integer solutions, (comment 1). If we solve for x we find: $x = (1068 - 278y)/117$. Now, according to Th. 1-2, we know that for one of the numbers $y \in \{0,1,2,3, \dots ,116\}$ the corresponding x will be integer. However, in practice, since 116 is a large number, we have to make many trials, (maximum 117 trials), in order to find the pair of integers that satisfy the equation. This might be very laborious and time consuming. For this reason, we present another method which facilitates the calculations, and can be applied when **the coefficients of the equation are large numbers**.

b) From the equality of the algorithmic division we have, (if we divide the numbers 1068 and 278 by 117):

$$1068 = 9 \cdot 117 + 15 \qquad 278 = 2 \cdot 117 + 44$$

Then,

$$x = \frac{1068 - 278y}{117} = \frac{9 \cdot 117 + 15 - (2 \cdot 117 + 44)y}{117} \Rightarrow$$

$$x = (9 - 2y) + \frac{15 - 44y}{117} \qquad (*)$$

Since x and y are integer numbers, eq. (*) implies that $(15 - 44y)/117$ **must be some integer**, say z, i.e.

$$\frac{15 - 44y}{117} = z \Rightarrow 44y + 117z = 15 \Rightarrow y = \frac{15 - 117z}{44}$$

or, since $117 = 2 \cdot 44 + 29$,

$$y = \frac{15 - (2 \cdot 44 + 29)z}{44} = -2z + \frac{15 - 29z}{44} \qquad (**)$$

Reasoning similarly, since y and z must be integers, the fraction $(15 - 29z)/44$ **must also be some integer**, say k, i.e.

$$\frac{15 - 29z}{44} = k \Rightarrow 29z + 44k = 15 \Rightarrow z = \frac{15 - 44k}{29}$$

or, since $44 = 1 \cdot 29 + 15$,

$$z = \frac{15 - (29 + 15)k}{29} \Rightarrow z = -k + \frac{15(1 - k)}{29} \qquad (***)$$

Since z and k are integers, an obvious solution for k, in eq. (***), is $\boldsymbol{k = 1}$, and this yields $z = -1$. Substituting this value of z in eq. (**) yields, $y = 3$, and then, from eq. (*), we find $x = 2$.

In summary, one pair of integer solutions of the original equation is $(x_0 = 2, y_0 = 3)$. The general solution of the equation is (see eq. (1-2-2)):

$$x = 2 + 278\lambda, \quad y = 3 - 117\lambda, \quad \lambda = 0, \pm 1, \pm 2, \pm 3, \ldots..$$

Example 1-2-5: Consider the Diophantine equation $(k^2 + k + 1)x + (k + 1)y = k$, where $k \in \mathbb{N}$. **a)** If d is the G.C.D of the coefficients $(k^2 + k + 1)$ and $(k + 1)$, show that $d = 1$, **b)** Show that the given equation has always integer solutions, and **c)** Find the general solution of the equation.

Solution: a) Let m be a common divisor of $(k^2 + k + 1)$ and $(k + 1)$. This means that $m \,/\, (k^2 + k + 1)$ and $m \,/\, (k + 1)$, and this implies that $m \,/\, (k^2 + k + 1)$ and $m \,/\, k(k + 1)$, (since k is a positive integer), and therefore m should also divide the difference $\{k^2 + k + 1\} - \{k(k + 1)\}$, i.e. $m \,/\, 1$, i.e. $\boldsymbol{m = 1}$. In other words, the only common divisor of $(k^2 + k + 1)$ and $(k + 1)$ is the number 1, i.e. $\boldsymbol{d = (k^2 + k + 1, k + 1) = 1}$.

b) By virtue of comment 1, the Diophantine equation admits integer solutions.

c) Solving the given equation for y, we find:

$$y = \frac{k - (k^2 + k + 1)x}{k + 1} = \frac{(k - x) - k(k + 1)x}{k + 1} = \frac{k - x}{k + 1} - kx \qquad (*)$$

Since x and y must be integers, the term $(k - x)/(k + 1)$, in eq. (*), must be integer as well, and this means that $\boldsymbol{x = -1}$. Then, the corresponding value of y, in eq. (*), is $\boldsymbol{y = 1 - k(-1) = 1 + k}$.

Thus, one integer solution of the original Diophantine equation is

$$x_0 = -1, \;\; y_0 = 1 + k \qquad (**)$$

The general solution of the equation is, (see eq. (1-2-2)):

$$x = -1 + (k + 1)\lambda, \;\; y = 1 + k - (k^2 + k + 1)\lambda, \quad \lambda = 0, \pm 1, \pm 2, \pm \pm 3, \dots$$

Example 1-2-6: Consider the Diophantine equation $a^2 x + b^2 y = 2ab$, where $a, b \in \mathbb{Z} - \{0\}$, and $a \neq b$. Show that this equation does not have positive, integer solutions.

Solution: Let us assume that the given equation has **a positive** solution (x_0, y_0), with $x_0 > 0$ **and** $y_0 > 0$. Then, the number $a^2 x_0 + b^2 y_0$ must be a positive number, i.e. $\boldsymbol{a^2 x_0 + b^2 y_0 > 0}$. Since x_0 and y_0 must be positive and integer numbers, $x_0 \geq 1$ **and** $y_0 \geq 1$. Thus $a^2 x_0 \geq a^2$ and $b^2 y_0 \geq b^2$, and therefore, $\boldsymbol{a^2 x_0 + b^2 y_0 \geq a^2 + b^2}$, or, since $a^2 x_0 + b^2 y_0 = 2ab$, it follows that $2ab \geq a^2 + b^2$, i.e. $0 \geq a^2 + b^2 - 2ab$, i.e. $\boldsymbol{0 \geq (a - b)^2}$, and this implies that $\boldsymbol{a = b}$, (since for $a \neq b$, $(a - b)^2 > 0$). However, by our hypothesis, $a \neq b$, and this shows that our original assumption that the equation has a positive solution cannot be true. In other words, the equation does not have integer positive solutions, and this completes the proof.

Example 1-2-7: Find the smallest positive integer a which when divided by 11 leaves remainder 3, while when divided by 19 leaves remainder 10.

Solution: From the equality of the algorithmic division we have:

$$\left.\begin{cases} a = 11x + 3 \\ \textbf{\textit{and}} \\ a = 19y + 10 \end{cases}\right\} \qquad (*)$$

where x is the quotient of the division $\{a \div 11\}$ and y is the quotient of the division $\{a \div 19\}$. Note that both x and y are positive and integer numbers. From eq. (*) it follows that:

$$11x + 3 = 19y + 10 \Longleftrightarrow 11x - 19y = 7 \qquad (**)$$

Thus, in order to find a, we need to find the positive and integer solutions of eq. (**). This equation does have integer solutions, since the numbers 11 and (-19) are relatively prime numbers. To find one solution, we solve equation (**) for x, i.e.

$$x = \frac{7 + 19y}{11} \qquad (***)$$

By successive trials we find that when $y = 6$, the corresponding value of x is, $x = 11$, and thus one solution of the equation is $(x_0 = 11, y_0 = 6)$. The general solution of the equation is:

$$x = 11 - 19k, \quad y = 6 - 11k, \quad k = 0, \pm 1, \pm 2, \pm 3, \dots. \qquad (****)$$

We see that when $k = 0, -1, -2, -3, \dots$, both x and y are positive, and thus, from eq. (*), a is positive. The smallest value of a is obtained for $k = 0$, and in this case, $x = 11, y = 6$ and form eq. (*) $a = \textbf{124}$, which is the sought for number.

PROBLEMS

1-2-1) Which ones of the following equations admit integer solutions?

$a)\ 2x + 9y = 10$ $b)\ 6x + 9y = 7$ $c)\ 8x + 4y = 12$
$d)\ 7x + 9y = 19$ $e)\ 4x - 5y = 10$ $f)\ 5x + 10y = 13$
$g)\ 8x + 24y = 49$ $h)\ 37x + 74y = 95$ $i)\ 65x + 75y = 100$

(Ans: (b), (f), (g), (h) do not admit integer solutions).

1-2-2) Find the integer solutions of the equations:

$$a)\ 18x - 21y = 9 \quad b)\ 7x + 5y = -6 \quad c)\ 13x + 19y = 25$$

1-2-3) Find the integer solutions of the equations:

$$a)\ 3x + 5y = 7 \quad b)\ 2x + 13y = 10 \quad c)\ 6x + 19y = 20$$

(Ans: a) $x = -1 + 5k, y = 2 - 3k,\ k \in \mathbb{Z}$, **b)** $x = 5 + 13k, y = -2k$,
$k \in \mathbb{Z}$, **c)** $x = -3 + 19k, y = 2 - 6k,\ k \in \mathbb{Z}$).

1-2-4) Find the integer and positive values of x that make the following expressions to be positive and integer:

$$a)\ \frac{x-1}{3} \qquad b)\ \frac{3x-10}{7} \qquad c)\ \frac{11x+8}{17}$$

1-2-5) Find a fraction which remains unaffected when its numerator is increased by 4 and its denominator is increased by 12,   **(Ans:** 1/3).

1-2-6) Express the fraction $77/65$ as the sum or the difference of two other fractions having denominators 5 and 13.

1-2-7) a) Show that the equation $(3k + 5)x + (5k + 8)y = k^2 + 2k$, always has integer roots, for any $k \in \mathbb{N}$, **b)** Show that one solution of the equation is $(x_0 = 2k, y_0 = -k)$, and then find its general solution.

(Ans: $x = 2k + (5k + 8)\lambda,\ y = -k - (3k + 5)\lambda, \lambda = 0, \pm1, \pm2, \pm3, ...)$.

Hint: a) Show that $(3k + 5, 5k + 8) = 1$. If d is a common divisor of $3k + 5$ and $5k + 8$, then, d must divide the number $5 \cdot (3k + 5) - 3 \cdot (5k + 8) = 1$, i.e. $d = 1$.

1-2-8) Consider the equation $ax + by = c$, where $a, b \in \mathbb{Z}$, $(a, b) = 1$, $ab > 0$ and $abc < 0$. Show that this equation does not have positive, integer solutions.

1-2-9) Find all the pairs of even integers that satisfy the equation $22x - 7y = 6$, (**Ans:** $x = 6 - 14k, y = 18 - 44k$, $k \in \mathbb{Z}$).

1-2-10) Find all the positive and integer solutions of the equation $11x - 3y = 37$.

(**Ans:** $x = 2 - 3k, y = -5 - 11k$, $k = -1, -2, -3, ...$).

1-3) Diophantine equations of the first degree in three unknowns, $(ax + by + cz = m)$

a) Let us consider the Diophantine equation $ax + by + cz = m$, where the coefficients a, b, c, m are integer numbers $(a, b, c, m \in \mathbb{Z})$, and $a \neq 0$, $b \neq 0$, $c \neq 0$.

Solution of this Diophantine equation is any ordered triad of integer numbers (x_0, y_0, z_0) that satisfies the equation, i.e. $ax_0 + by_0 + cz_0 = m$.

Solving the Diophantine equation means **to find all its integer solutions**.

For example, the numbers $x_0 = 1, y_0 = 2, z_0 = 4$ constitute a solution of the equation $2x + 3y + 7z = 36$, since $2 \cdot 1 + 3 \cdot 2 + 7 \cdot 4 = 36$. Another solution is $(x_0 = 11, y_0 = 7, z_0 = -1)$, since $2 \cdot 11 + 3 \cdot 7 + 7 \cdot (-1) = 36$, etc.

b) The following theorems are important in solving $ax + by + cz = m$.

Theorem 1-4: If a, b and c have a common divisor q, which does not divide m, then the equation $ax + by + cz = m$ does not have any integer solutions.

Proof: Similar to the proof of Theorem 1-1.

Theorem 1-5: Let d be the Greatest Common Divisor of a, b and c, i.e. let $d = (a, b, c)$. Then, the Diophantine equation $ax + by + cz = m$ has an integer solution, if and only if, d divides m.

Proof: Let us call f the G.C.D of b and c, i.e. $f = (b, c)$. Then, the equation $ax + fw = m$, (x and w are now the unknowns), does have integer solutions, since the G.C.D of a and f, which is actually the G.C.D of a, b and c, divides m, (by assumption). Let (x_0, w_0) be an integer solution, i.e.

$$ax_0 + fw_0 = m \qquad (*)$$

Let us now consider the equation $by + cz = fw_0$. This equation admits an integer solution since $(b, c) = f$ divides the right term fw_0, (see Th. 1-2). Thus, there exists a pair of integers (y_0, z_0) such that

$$by_0 + cz_0 = fw_0 \qquad (**)$$

Adding equations (*) and (**) yields:

$$ax_0 + by_0 + cz_0 = m \qquad (***)$$

and this shows that (x_0, y_0, z_0) **is a solution of the Diophantine equation** $ax + by + cz = m$.

c) As we shall see, the general solution of the Diophantine equation $ax + by + cz = m$ is expressed in terms of **two integer parameters**, (as opposed to the general solution of $ax + by = c$, which is expressed in terms of **one integer parameter**).

d) Solving the Diophantine equation $ax + by + cz = m$.

We start with the simplest case where **one of the coefficients a or b or c is 1**, let us, for example, assume $a = 1$. In this case the Diophantine equation is $x + by + cz = m$, and solving for x we find:

$$x = m - by - cz \qquad (1 - 3 - 1)$$

The **general solution** of this equation is obviously given by the formulas:

$$\left. \begin{cases} y = k & (integer) \\ z = \lambda & (integer) \\ x = m - bk - c\lambda \end{cases} \right\} \quad where: \quad \begin{aligned} k &= 0, \pm1, \pm2, \pm3, ... \\ \lambda &= 0, \pm1, \pm2, \pm3, ... \end{aligned} \quad (1-3-2)$$

As we see, the general solution contains two integer parameters, k and λ.

For example, to solve $x + 3y - 5z = 10$, we write the equation in the form $x = 10 - 3y + 5z$, and its general solution is:

$$y = k \in \mathbb{Z}$$
$$z = \lambda \in \mathbb{Z}$$
$$x = 10 - 3k + 5\lambda$$

For $k = 1$ and $\lambda = 1$, we find one solution $(x_0 = 12, y_0 = 1, z_0 = 1)$, for $k = 2$ and $\lambda = -1$, we find another solution $(x_0 = -1, y_0 = 2, z_0 = -1)$, etc.

In case **none of the coefficients is 1**, then, we try to reduce the given equation to an equivalent one, having smaller coefficients, and then again reduce this equation to another one, with even smaller coefficients, etc. until we reach to an equation having one of its coefficients equal to 1, and then solve as described previously. The method is illustrated in the following examples.

Example 1-3-1: Solve the Diophantine equation: $2x + 3y - 4z = 10$.

Solution: Solving for x we find:

$$x = \frac{10 - 3y + 4z}{2} = 5 + 2z - \frac{3y}{2} \quad (*)$$

Since x and $(5 + 2z)$ are integers, **the fraction $3y/2$ must be an integer as well**, and this implies that $y = 2k$, with $k \in \mathbb{Z}$. Thus, the general solution of the given equation is:

$$\left. \begin{cases} x = 5 + 2\lambda - 3k \\ y = 2k \\ z = \lambda \end{cases} \right\} \quad where: \quad k, \lambda = 0, \pm1, \pm2, \pm3, ...$$

For example, for $k = 2, \lambda = 2$ we find one solution: $(x = 3, y = 4, z = 2)$. For $k = -1, \lambda = 3$ we find another solution: $(x = 14, y = -2, z = 3)$, etc.

Example 1-3-2: Solve the Diophantine equation: $6x + 22y + 15z = 77$.

Solution: The equation does have integer solutions, since $(6,22,15) = 1$, i.e. the G.C.D of $6, 22$ and 15 is 1, (see Th. 1-5). Solving this equation for x we find:

$$x = \frac{77 - 22y - 15z}{6} \qquad (*)$$

Dividing $77, 22$ and 15 by 6 we find, (equality of algorithmic division):

$$77 = 6 \cdot 12 + 5 \quad 22 = 6 \cdot 3 + 4 \quad 15 = 6 \cdot 2 + 3$$

Equation (*) becomes,

$$x = \frac{6 \cdot 12 + 5 - (6 \cdot 3 + 4)y - (6 \cdot 2 + 3)z}{6} \Longrightarrow$$

$$x = 12 - 3y - 2z + \frac{5 - 4y - 3z}{6} \qquad (**)$$

Since x, y and z are integers, **the fraction $(5 - 4y - 3z)/6$ must likewise be some integer**, say n, i.e.

$$\frac{5 - 4y - 3z}{6} = n \Longrightarrow 4y + 3z + 6n = 5 \Longrightarrow z = \frac{5 - 4y - 6n}{3} \Longrightarrow$$

$$z = \frac{(3 + 2) - (3 + 1)y - 6n}{3} = 1 - y - 2n + \frac{2 - y}{3} \qquad (***)$$

Again, since z, y and n are integers, **the fraction $(2 - y)/3$ must be some integer**, say m, i.e.

$$\frac{2 - y}{3} = m \Longrightarrow y + 3m = 2 \qquad (****)$$

This is a first degree Diophantine equation in two unknowns m and y. The general solution of eq. (****) is obvious, $m \in \mathbb{Z}, y = (2 - 3m) \in \mathbb{Z}$. Then, from eq. (***) we find $z = -1 + 4m - 2n$, and finally, from eq. (**) we find $x = 8 + m + 5n$. Summarizing, **the general solution** of the given equation is:

$$\left\{\begin{array}{l} x = 8 + m + 5n \\ y = 2 - 3m \\ z = -1 + 4m - 2n \end{array}\right\} \qquad where: \ m, n = 0, \pm 1, \pm 2, \pm 3, \dots ..$$

For example, for $m = 1, n = 2$ we find, $(x = 19, y = -1, z = -1)$, for $m = 2, n = 3$ we find another solution $(x = 25, y = -4, z = 1)$, etc.

Example 1-3-3: Solve the Diophantine equation: $2|x| + 3|y| + 5|z| = 25$.

Solution: Let us, for simplicity, set: $a = |x|, b = |y|, c = |z|$. The given equation becomes,

$$2a + 3b + 5c = 25, \ \ with \ \ a, b, c \ integers \ and \ a \ge 0, b \ge 0, c \ge 0 \qquad (*)$$

Solving for a we find:

$$a = \frac{25 - 3b - 5c}{2} = \frac{(2 \cdot 12 + 1) - (2 + 1)b - (2 \cdot 2 + 1)c}{2} \Longrightarrow$$

$$a = 12 - b - 2c + \frac{1 - b - c}{2} \qquad (**)$$

Since a and $(12 - b - 2c)$ are integers, **the fraction $(1 - b - c)/2$ must necessarily be some integer**, say k, i.e.

$$\frac{1 - b - c}{2} = k \Longrightarrow b + c + 2k = 1 \Longrightarrow b = 1 - 2k - c \qquad (***)$$

The general solution of eq. (***) is: $c = \lambda \ (integer), \ b = 1 - 2k - \lambda$, and then, from eq. (**), we find, $a = 11 + 3k + 3\lambda$.

Summarizing, the general solution of equation $2a + 3b + 5c = 25$ is:

$$\left\{\begin{array}{l} a = 11 + 3k - \lambda \\ b = 1 - 2k - \lambda \\ c = \lambda \end{array}\right\} \qquad (****)$$

where k and λ are integers which make $a \ge 0$, and $b \ge 0$, and $c \ge 0$, simultaneously.

From the third equation, in eq. (****), it follows that $c = \lambda \ge 0$. Then, from the second equation it follows,

$$b = 1 - 2k - \lambda \geq 0 \Longrightarrow 1 - 2k \geq \lambda \geq 0 \Longrightarrow k \leq 1/2$$

while from the first equation, we have,

$$a = 11 + 3k - \lambda \geq 0 \Longrightarrow 11 + 3k \geq \lambda \geq 0 \Longrightarrow k \geq -\ 11/3 \cong -3.66\ ....$$

Thus, $-3.66\ ... \leq k \leq 1/2$, and since k is an integer number, **the allowed values of k are: $-3, -2, -1, 0$.**

a) For $k = 0$, the equations in eq. (****) become:

$$\left\{ \begin{array}{l} a = 11 - \lambda \\ b = 1 - \lambda \\ c = \lambda \end{array} \right\} \qquad (*****)$$

The allowed integer values of λ are the ones for which the three inequalities in eq. (*****) are satisfied simultaneously, i.e. $c = \lambda \geq 0$, $b = 1 - \lambda \geq 0$, i.e. $\lambda \leq 1$, and $a = 11 - \lambda \geq 0$, i.e. $\lambda \leq 11$.

The integer values of λ that satisfy, **simultaneously** the inequalities $\lambda \geq 0$ **and $\lambda \leq 1$ and $\lambda \leq 11$** are: $\lambda = 0$ or $\lambda = 1$.

Working similarly, we find, (for details see Pr. 1-3-7):

b) For $k = -1$, $\lambda = 0,1,2,3$.

c) For $k = -2$, $\lambda = 0,1,2,3,4,5$.

d) For $k = -3$, $\lambda = 0,1,2$.

The results we have obtained are summarized in the following Table 1:

Table 1: Allowed values of pairs (k, λ)

	$k = -3$	$k = -2$	$k = -1$	$k = 0$
λ	$0, 1, 2$	$0, 1, 2, 3, 4, 5$	$0, 1, 2, 3$	$0, 1$

Thus, for example, for $k = -3, \lambda = 1$, we find (from eq. (****)), $a = 1, b = 6, c = 1$, i.e. $|x| = 1, |y| = 6, |z| = 1$, i.e. $x = \pm 1, y = \pm 6, z = \pm 1$.

For another allowed pair of k and λ, say $k = -2, \lambda = 3$, we find, (from equation (****)):

$a = 2, b = 2, c = 3$, i.e. $|x| = 2, |y| = 2, |z| = 3$, and thus $x = \pm 2, y = \pm 2, z = \pm 3$, etc.

PROBLEMS

1-3-1) Show that the following Diophantine equations do **not** have integer solutions:

$$2x + 4y + 6z = 9, \quad 7x + 21y + 14z = 50, \quad 5x - 10y + 20z = 39$$

1-3-2) Solve the Diophantine equation: $3x + 5y + 7z = 89$.

1-3-3) Solve the Diophantine equation: $2x + 5y + 7z = 111$.

(**Ans:** $x = 53 + 5k - \lambda, \; y = 1 - 2k - \lambda, \; z = \lambda, \; k, \lambda = 0, \pm 1, \pm 2, \pm 3, ...$).

1-3-4) Show that the equation $12|x| + 20|y| + 15|z| = 60$ is not satisfied by any triad of natural numbers.

Hint: The possible values of x that may satisfy the equation are: $1, 2, 3, 4$. Show that each one of these values leads to an equation in $|y|$ and $|z|$, which does not admit integer solutions.

1-3-5) Solve the Diophantine equation: $4x + 3y + 7z = 11$.

(**Ans:** $x = 2 - 3k - \lambda, y = 1 + 4k - \lambda, z = \lambda, \; k, \lambda \in \mathbb{Z}$).

1-3-6) Solve the Diophantine equations:

$$2|x| + 3|y| + |z| = 7, \quad 11x + 7y + 7z = 13, \quad x + 2y + 3z = 5$$

1-3-7) Verify the values for k and λ, listed in Table 1, in Ex. 1-3-3.

1-3-8) Find all the integer solutions (x_0, y_0, z_0) of $2x + 7y + 11z = 20$, that satisfy the relation $x_0 + y_0 + z_0 = 10$.

(**Ans:** $x = 10 - 4k, y = 9k, z = -5k, \; k = 0, \pm 1, \pm 2, \pm 3, ...$).

1-4) Diophantine equations of the first degree in more than three unknowns

a) We may have a Diophantine equation, of the first degree in four, or five or six, etc. unknowns. For example, the equation: $ax + by + cz + dw = f$, where a, b, c, d, f are integers, is a first degree equation in the four unknowns x, y, z, w. Solving this Diophantine equation means to find **all the integers** (x_0, y_0, z_0, w_0) that satisfy the equation, i.e., $ax_0 + by_0 + cz_0 + dw_0 = f$. As we shall see, **the general solution of this equation contains three integer parameters.**

b) All the theorems for a first degree equation in three unknowns remain valid in this case as well. For example, if the coefficients a, b, c, d have a common divisor which does not divide the constant term f, then the equation does **not** have integer solutions.

c) The method of solution of a Diophantine equation in four unknowns is very much similar to the method of solution of an equation in three unknowns, and is illustrated hereunder by means of a few examples.

Example 1-4-1: Solve the Diophantine equation: $x - 3y + 5z + 9w = 15$.

Solution: Since the coefficient of x is 1, we solve the equation for x and find:

$$x = 15 + 3y - 5z - 9w$$

The obvious solution of this equation is

$$\left\{ \begin{array}{c} x = 15 + 3k - 5\lambda - 9\mu \\ y = k \\ z = \lambda \\ w = \mu \end{array} \right\} \quad k, \lambda, \mu = 0, \pm1, \pm2, \pm3, \ldots\ldots$$

As we see, the solution contains three integer parameters, k, λ and μ. For example, for $k = 1, \lambda = 1, \mu = 2$ we find one solution $(-5,1,1,2)$, for $k = -1, \lambda = 3, \mu = 1$ we find another solution $(-12, -1,3,1)$, etc.

Example 1-4-2: Solve the Diophantine equation: $3x - 2y + 5z + 6w = 80$.

Solution: The smallest (in absolute value) coefficient is 2, (the coefficient of y). Solving the given equation for y we find:

$$y = \frac{-80 + 3x + 5z + 6w}{2} \implies$$

$$y = \frac{-2 \cdot 40 + (2 + 1)x + (2 \cdot 2 + 1)z + (2 \cdot 3)w}{2} \implies$$

$$y = -40 + x + 2z + 3w + \frac{x + z}{2} \qquad (*)$$

Since y and $(-40 + x + 2z + 3w)$ are integers, **the fraction $(x + z)/2$ in eq. (*) must necessarily be some integer**, say k, i.e.

$$\frac{x + z}{2} = k \implies x + z = 2k \implies x = 2k - z \qquad (**)$$

An obvious solution of eq. (**) is, $z = \lambda$ (integer), $x = 2k - \lambda$. Substituting in eq. (*) and setting $w = \mu$, (integer), we find: $y = -40 + 4k - \lambda + 5\mu$.

Summarizing, the general solution of the equation is:

$$\left\{ \begin{array}{c} x = 2k - \lambda \\ y = -40 + 3k + \lambda + 3\mu \\ z = \lambda \\ w = \mu \end{array} \right\}, \quad k, \lambda, \mu = 0, \pm 1, \pm 2, \pm 3, \ldots \qquad (***)$$

For example, for $k = 1, \lambda = -2, \mu = -1$ we find:
$(x = 4, y = -42, z = -2, w = -1)$. This is one solution. If $k = 2, \lambda = -1$ and $\mu = 3$ we find another solution $(x = 4, y = -31, z = -1, w = 3)$, etc.

Example 1-4-3: Find all the integer solutions (x_0, y_0, z_0, w_0) of the Diophantine equation in Example 1-4-2, which satisfy the relation $x_0 + y_0 + z_0 + w_0 = -43$.

Solution: The general solution of the Diophantine equation is given by the formula (***), in the previous example, and since $x_0 + y_0 + z_0 + w_0 = -43$, we must have:

$$(2k - \lambda) + (-40 + 3k + \lambda + 3\mu) + \lambda + \mu = -43 \implies$$

$$5k + \lambda + 4\mu = -3 \Longrightarrow \lambda = -3 - 5k - 4\mu \qquad (*)$$

Substituting this expression of λ into the equations in formula (***), in Ex. 1-4-2, we find:

$$\begin{cases} x = 7k + 4\mu + 3 \\ y = -43 - 2k - \mu \\ z = -3 - 5k - 4\mu \\ w = \mu \end{cases}, \quad k, \mu = 0, \pm 1, \pm 2, \pm 3, ... \qquad (**)$$

For example, for $\mu = -1, k = 2$, we find:
$x = 13, y = -46, z = -9, w = -1$.

PROBLEMS

1-4-1) Show that the following equations do not admit integer solutions:

$$2x + 4y + 6z + 10w = 17, \quad 3x - 9y + 21z - 36w = 112$$

1-4-2) Solve the Diophantine equations:

$$3x + 5y + 7z - w = 10, \quad 4x + 6y + 12z + 16w = 50$$

1-4-3) Solve the Diophantine equations:

$$a)\ 3x + 5y + 7z + 9w = 27, \quad b)\ 5x - 3y - 2z + 11w = 20$$

(**Ans: a)** $x = 9 - 7k + 3\lambda - 3\mu, \ y = \lambda, \ z = 3k - 2\lambda, \ w = \mu,$

b) $x = 2k + \lambda - \mu, \ y = \lambda, \ z = -10 + 5k + \lambda + 3\mu, \ w = \mu$).

1-4-4) Find all the integer solutions (x_0, y_0, z_0, w_0) of the first Diophantine equation in Pr. 1-4-3, which satisfy the relation $2x_0 + 3y_0 = z_0 + 5w_0 + 7$.

(**Ans:** $x = 6 - 26\rho, \ y = -1 + 17\rho + \mu, \ z = 2 - \rho - 2\mu, \ w = \mu$, where $\rho, \mu = 0, \pm 1, \pm 2, \pm 3, ...$).

1-4-5) Find the integer numbers x, y, z, w which satisfy the equation:
$$|x + 3| + |y - 10| + |z - 7| + |w + 5| = 0$$

(**Ans:** $x = -3, y = 10, z = 7, w = -5$).

CHAPTER 2: DIOPHANTINE SYSTEMS OF LINEAR EQUATIONS

2-1) Diophantine systems of two, first degree equations, with three unknowns

a) Let us consider a system of two linear equations with three unknowns, of the general form

$$\begin{cases} ax + by + cz = f \\ Ax + By + Cz = F \end{cases}$$

This system is called "**a Diophantine system**" if all the coefficients (a, b, c, f, A, B, C, F) are integers and if only integer solutions are allowed.

An ordered triad of integers (x_0, y_0, z_0) is called **a solution** of the system, if they satisfy both equations, **simultaneously**, i.e. if

$$\{ax_0 + by_0 + cz_0 = f \quad \textbf{and} \quad Ax_0 + By_0 + Cz_0 = F\}$$

Solving the Diophantine system means to find **all its integer solutions**.

If one equation of the system does not have integer solutions, then the system does not have integer solutions. For example, the system

$$\begin{cases} 2x + 4y + 8z = 19 \\ 3x + 6y + 9z = 21 \end{cases}$$

does not have integer solutions, because the first equation does not have integer solutions (why?).

b) The main idea to solve a Diophantine system is to eliminate one of the unknowns and thus obtain one equation with two unknowns. The elimination is achieved by a proper linear combination of the two equations.

The following examples illustrate the method.

Example 2-1-1: Solve the Diophantine system

$$\begin{cases} 2x + 5y - 11z = 1 \\ x - 12y + 7z = 2 \end{cases}$$

Solution: Solving the second equation for x we find, $x = 2 + 12y - 7z$, and substituting into the first equation yields, $25z - 29y = 3$, (thus far we have eliminated the unknown x between the two equations and have obtained one equation with two unknowns). This is a Diophantine equation, in two unknowns z and y, whose general solution is found to be

$$\{y = 18 - 25k \qquad z = 21 - 29k\} \quad k = 0, \pm 1, \pm 2, \pm 3, \ldots \ldots \qquad (*)$$

Then, since $x = 2 + 12y - 7z$, we find, $x = 71 - 97k$.

Example 2-1-2: Solve the Diophantine system

$$\begin{cases} (E_1) \quad 3x + 4y + 5z = 13 \\ (E_2) \quad 6x + 6y - 35z = 29 \end{cases}$$

Solution: If we call (E_1) the first equation and (E_2) the second, then, if we consider **the linear combination** $2 \cdot (E_1) - (E_2)$, (i.e. multiply the first equation by 2 and then subtract the second one), we find:

$$6x + 8y + 10z - 6x - 6y + 35z = 26 - 29 \Rightarrow$$

$$2y + 45z = -3 \qquad (*)$$

We have thus eliminated x between the two original equations. Equation $(*)$ is a first degree equation in two unknowns, y and z. Solving for y we find

$$y = -\frac{45z + 3}{2}$$

As we know, (see section 1-2), for one $z \in \{0,1\}$ the corresponding y will be integer. Here we find easily that for $z = 1$ the corresponding $y = -24$. The general solution of eq. $(*)$ is

$$\{y = -24 + 45k, \qquad z = 1 - 2k\}, \ k \in \mathbb{Z} \qquad (**)$$

The thus found expressions of y and z must also satisfy one of the original equations, for example the first one, so if we replace the expressions of y and z, (from eq. (**)) into the first equation (E_1) we have:

$$3x + 4(-24 + 45k) + 5(1 - 2k) = 13 \Longrightarrow 3x + 170k = 104 \quad (***)$$

Equation (***) is another Diophantine equation for x and k. In particular, note that k cannot just be any integer, on the contrary, the general expression of the allowed values of k will be obtained from the solution of eq. (***). Solving for x we find:

$$x = \frac{104 - 170k}{3}$$

We note that when $k = 1$ the corresponding $x = -22$, and the general solution of eq. (***) is

$$\{x = -22 + 170\lambda \qquad k = 1 - 3\lambda\} \quad \lambda \in \mathbb{Z}$$

Substituting this expression of k, (in terms of λ), into the two equations in formula (**) we find the general solution of the system:

$$\begin{cases} x = -22 + 170\lambda \\ y = 21 - 135\lambda \\ z = -1 + 6\lambda \end{cases} \quad \lambda = 0, \pm 1, \pm 2, \pm 3, \ldots..$$

Example 2-1-3: Find the positive solutions of the Diophantine system

$$\begin{cases} (E_1) \quad 3x + 2y + 3z = 250 \\ (E_2) \quad 9x - 4y + 5z = 170 \end{cases}$$

Solution: In order to eliminate x between the two equations we consider the linear combination $3 \cdot (E_1) - (E_2)$:

$$3 \cdot (3x + 2y + 3z) - (9x - 4y + 5z) = 3 \cdot 250 - 170 \Longrightarrow$$

$$10y + 4z = 580 \Longrightarrow 5y + 2z = 290 \qquad (*)$$

Equation (*) is a Diophantine equation in two unknowns, y and z. When $y = 0$ the corresponding $z = 145$. Thus, the general solution of eq. (*) is

$$\{y = 2k \qquad z = 145 - 5k\} \quad k \in \mathbb{Z} \qquad\qquad (**)$$

Substituting these values of y and z in (E_1) we find:

$$3x - 11k = -185 \qquad\qquad (***)$$

Equation (***) is another Diophantine equation with unknowns x and k. It follows that $x = (11k - 185)/3$, and when $k = 1, x = -58$, and therefore the general solution of eq. (***) is:

$$\{x = -58 - 11\lambda \qquad k = 1 - 3\lambda\} \qquad\qquad (****)$$

Substituting the allowed values of k, (from eq. (****)), in eq. (**), we find the general solution of the original system:

$$\begin{cases} x = -58 - 11\lambda \\ \quad y = 2 - 6\lambda \\ z = 140 + 15\lambda \end{cases} \lambda = 0, \pm1, \pm2, \pm3, \ldots\ldots \qquad (*****)$$

To find the positive solutions means to find the values of λ for which $x > 0$ and $y > 0$ and $z > 0$.

$$x > 0 \implies -58 - 11\lambda > 0 \implies 58 + 11\lambda < 0 \implies \lambda < -58/11 \cong -5.27 \ldots$$

$$y > 0 \implies 2 - 6\lambda > 0 \implies \lambda < 2/6 \cong 0.33 \ldots$$

$$z > 0 \implies 140 + 15\lambda > 0 \implies \lambda > -140/15 \cong -9.33 \ldots$$

These three inequalities are satisfied **simultaneously** for four values of λ, i.e. $\lambda = -6, -7, -8, -9$, and then, from eq. (*****) we find:

For $\lambda = -6$: $(x = 8, y = 38, z = 50)$.

For $\lambda = -7$: $(x = 19, y = 44, z = 35)$.

For $\lambda = -8$: $(x = 30, y = 50, z = 20)$.

For $\lambda = -9$: $(x = 41, y = 56, z = 5)$.

These are the only positive and integer solution of the given system.

Comment: So far, we have solved linear Diophantine systems using the **method of elimination** of one unknown. This is not the only method. Another method could be, for example, **to solve the first equation of the system**, as described in section 1-3, and then, use the second equation to obtain the general solution. In the next example, we solve the system of the previous example, following this approach.

Example 2-1-4: Find the positive solutions of the Diophantine system

$$\begin{cases} (E_1) & 3x + 2y + 3z = 250 \\ (E_2) & 9x - 4y + 5z = 170 \end{cases}$$

Solution: Since the smallest coefficient of the unknowns in eq. (E_1) is 2, (the coefficient of y), we solve the equation for y:

$$y = \frac{250 - 3x - 3z}{2} = \frac{2 \cdot 125 - (2+1)x - (2+1)z}{2} \Longrightarrow$$

$$y = 125 - x - z - \frac{x+z}{2} \qquad (*)$$

Since y and $(125 - x - z)$ are integers, the fraction $(x + z)/2$ must likewise be some integer, say k, i.e.

$$\frac{x+z}{2} = k \Longrightarrow x + z = 2k \Longrightarrow x = 2k - z$$

The general solution of this equation is:

$$\{z = \lambda \qquad x = 2k - \lambda\} \qquad k, \lambda = 0, \pm 1, \pm 2, \pm 3, \dots$$

The corresponding y is now found from eq. (*), i.e.

$$y = 125 - (2k - \lambda) - \lambda - k = 125 - 3k$$

Thus, the general solution of eq. (E_1) is:

$$\begin{cases} x = 2k - \lambda \\ y = 125 - 3k \\ z = \lambda \end{cases} \qquad k, \lambda = 0, \pm 1, \pm 2, \pm 3, \dots. \qquad (**)$$

In view of eq. (**), equation (E_2) yields:

$$9 \cdot (2k - \lambda) - 4 \cdot (125 - 3k) + 5\lambda = 170 \Longrightarrow 30k - 4\lambda = 670 \Longrightarrow$$

$$15k - 2\lambda = 335 \Longrightarrow \lambda = \frac{15k - 335}{2}$$

This is another Diophantine equation, with two unknowns k and λ. For $k = 1$, the corresponding $\lambda = -160$. Thus the general solution is

$$k = 1 - 2\mu, \quad \lambda = -160 - 15\mu, \quad \mu = 0, \pm 1, \pm 2, \pm \pm 3, \ldots \ldots \quad (***)$$

Substituting these expressions of k and λ, (in terms of μ), in eq. (**) we find:

$$\left.\begin{cases} x = 11\mu + 162 \\ y = 6\mu + 122 \\ z = -15\mu - 160 \end{cases}\right\} \quad \mu = 0, \pm 1, \pm 2, \pm \pm 3, \ldots \ldots \quad (****)$$

The positive solutions of the system are found from the inequalities: $x > 0$ **and** $y > 0$ **and** $z > 0$, i.e.

$$11\mu + 162 > 0 \quad \boldsymbol{and} \quad 6\mu + 122 > 0 \quad \boldsymbol{and} \quad -15\mu - 160 > 0$$

The first inequality is satisfied for $\mu > -162/11 \cong -14.72 \ldots$, the second inequality is satisfied for $\mu > -122/6 \cong -20.33 \ldots$, and the third inequality is satisfied for $\mu < -160/15 \cong -10.66 \ldots$ The three inequalities are satisfied simultaneously only for four values of μ, i.e. for $\mu = -11, -12, -13, -14$. From eq. (****) we find:

For $\mu = -11$: $(x = 41, y = 56, z = 5)$.

For $\mu = -12$: $(x = 30, y = 50, z = 20)$.

For $\mu = -13$: $(x = 19, y = 44, z = 35)$.

For $\mu = -14$: $(x = 8, y = 38, z = 50)$.

The results are identical to the ones found in the previous example.

Example 2-1-5: For the Diophantine system in Example 2-1-4, find the solutions which satisfy the inequality, $2x + 3y + 5z > 10$.

Solution

The general solution of the system is given by eq. (****), in Ex. 2-1-4. We want

$$2x + 3y + 5z > 10 \Rightarrow$$

$$2(11\mu + 162) + 3(6\mu + 122) + 5(-15\mu - 160) > 10 \Rightarrow$$

$$-35\mu - 110 > 10 \Rightarrow -35\mu > 120 \Rightarrow \mu < -120/35 \cong -3.4 \dots$$

Thus, the sought for solutions are:

$$\begin{cases} x = 11\mu + 162 \\ y = 6\mu + 122 \\ z = -15\mu - 160 \end{cases} \quad \mu = -4, -5, -6, \dots$$

PROBLEMS

2-1-1) Solve the Diophantine systems:

$$\begin{cases} 3x + 2y - 7z = 12 \\ 4x + 3y - 2z = 1 \end{cases} \qquad \begin{cases} 2x - 3y + 4z = 19 \\ 3x + 5y - 2z = -5 \end{cases}$$

(Ans: a) $x = 34 + 17k, y = -45 - 22k, z = k, \ k \in \mathbb{Z}$

b) $x = 2 - 14\lambda, y = -1 + 16\lambda, z = 3 + 19\lambda, \quad \lambda \in \mathbb{Z}$).

2-1-2) Solve the Diophantine systems:

$$\begin{cases} 5x + 12y + 2z = 49 \\ 6x - 5y - 2z = -17 \end{cases} \qquad \begin{cases} 5x + 6y + 7z = 18 \\ 2x + 3y + 4z = 9 \end{cases}$$

2-1-3) Find the integer numbers which satisfy the following relations:

$$\{3x - 11y + 2z = 13, \qquad 2x + y + 4z = 25\}$$

(Ans: $x = 6 - 46k, \ y = 1 - 8k, \ z = 3 + 25k, \ k = 0, \pm 1, \pm 2, \pm 3$).

2-1-4) Find the positive solution of the Diophantine system in Pr. 2-1-3.

2-1-5) Solve the Diophantine system:

$$\left\{\begin{array}{l} 3x - 5y + 6z = 11 \\ x + 2y - 7z = 16 \end{array}\right\}$$

(**Ans:** $x = 26 - 23\lambda, y = 23 - 27\lambda, z = 8 - 11\lambda, \quad \lambda = 0, \pm1, \pm2, \pm3, ...$).

2-1-6) Show that the system $\{x - y + z = 12, \quad 2x + y - z = 13\}$ does not have integer solutions.

2-1-7) Find the positive and integer solutions of the systems:

$$\left\{\begin{array}{l} 13x + 11z = 103 \\ 7z - 5y = 4 \end{array}\right\} \qquad \left\{\begin{array}{l} 5x + 4y + z = 272 \\ 8x + 9y + 3z = 656 \end{array}\right\}$$

(**Ans: a)** General solution: $x = 2 - 55k, y = 9 + 91k, z = 7 + 65k, k \in \mathbb{Z}$, positive solution, $(x = 2, y = 9, z = 7)$,

b) General solution: $x = 1 + 3k, \ y = 51 - 7k, z = 63 + 13k, \quad k \in \mathbb{Z}$. The positive solutions are obtained for, $k = 0, 1, 2, 3, 4, 5, 6, 7$).

2-2) Diophantine systems of $(n - 1)$, first degree equations, with n unknowns, $(n = 3, 4, 5, ...)$

The case $n = 3$ was studied in the previous section, (two equations in three unknowns). Working similarly, we may solve Diophantine systems of 3 equations with 4 unknowns, or, 4 equations with 5 unknowns, etc.

The method of solution is described below by means of some illustrative examples.

Example 2-2-1: Solve the Diophantine system:

$$\left\{\begin{array}{l} 5x + 2y + 3z + w = 9 \\ 2x - y + 5z + 3w = 4 \\ 3x + 5y + 4z + 6w = 27 \end{array}\right\}$$

Solution: In this case we have a system of three equations with four unknowns. Since the coefficient of w is 1, (in the first equation), we solve this equation for w and find:

$$w = 9 - 5x - 2y - 3z \qquad (*)$$

The general solution of eq. (*) is:

$$x = k, \qquad y = \lambda, \qquad z = \mu, \qquad w = 9 - 5k - 2\lambda - 3\mu, \quad k, \lambda, \mu \in \mathbb{Z} \quad (**)$$

Substituting x, y, z, w from eq. (**) into the second and the third equation of the original system, we find:

$$\left\{ \begin{array}{l} 13k + 7\lambda + 4\mu = 23 \\ 27k + 7\lambda + 14\mu = 27 \end{array} \right\} \qquad (***)$$

The system in eq. (***) is a system of two equations with three unknowns, (k, λ, μ), and can be solved as described in section 2-1.

Subtracting the first from the second equation, leads to the elimination of the unknown λ, i.e.

$$14k + 10\mu = 4, \quad or, \quad 7k + 5\mu = 2 \qquad (****)$$

Equation (****) is a Diophantine equation with unknowns k and μ, and solving for μ we find:

$$\mu = \frac{2 - 7k}{5}$$

For $\boldsymbol{k = 1}$, the corresponding $\boldsymbol{\mu = -1}$, and thus, thus general solution of eq. (****) is

$$k = 1 + 5m, \quad \mu = -1 - 7m, \qquad m \in \mathbb{Z} \qquad (*****)$$

Then, from one of the two equations in (***), say, for example, the first equation, we find:

$$7\lambda = 23 - 13k - 4\mu = 23 - 13(1 + 5m) - 4(-1 - 7m) \Rightarrow$$

$$7\lambda = 14 - 37m \Rightarrow 7\lambda + 37m = 14 \qquad (******)$$

This is another Diophantine equation, for λ and m. Solving for λ we find:

$$\lambda = \frac{14 - 37m}{7}$$

When $m = 0$ the corresponding $\lambda = 2$, and thus the general solution of equation (******) is:

$$\lambda = 2 + 37b, \quad m = -7b, \quad b = 0, \pm1, \pm2, \pm3, \dots \dots$$

Then, from eq. (*****) we find, $k = 1 - 35b$, $\mu = -1 + 49b$, and substituting the expressions of k, λ and μ, (in terms of b), in eq. (**) we find the general solution of the given system, (in terms of the integer parameter b):

$$\left.\begin{cases} x = 1 - 35b \\ y = 2 + 37b \\ z = -1 + 49b \\ w = 3 - 46b \end{cases}\right\} \quad b = 0, \pm1, \pm2, \pm3, \dots.$$

Example 2-2-2: Find two positive integers such that, the first one to be a multiple of 7 increased by 1, the second one to be a multiple of 9 increased by 7, and their sum to be 100.

Solution: Let x and y be the sought for integers. Since x is a multiple of 7 increased by 1, we may set $x = 7z + 1$, $z \in \mathbb{N}$, and since y is a multiple of 9 increased by 7, we may set $y = 9w + 7$, $w \in \mathbb{N}$. Thus, the numbers x and y shall be determined from the equations:

$$\{x = 7z + 1, \quad y = 9w + 7, \quad x + y = 100\} \quad x, y, z, w = 1, 2, 3, \dots \quad (*)$$

The system in eq. (*) is a Diophantine system, three equations in four unknowns. All the unknowns, (x, y, z, w), must be positive integers.

The general solution of the first equation is $z = k \in \mathbb{N}$, $x = 7k + 1$, while the solution of the second equation is $w = \lambda \in \mathbb{N}$, $y = 9\lambda + 7$. The third equation becomes:

$$(7k + 1) + (9\lambda + 7) = 100 \implies 7k + 9\lambda = 92 \quad (**)$$

This is a Diophantine equation in the unknowns k and λ. We easily find that when $\lambda = 4$, the corresponding $k = 8$, and thus the general solution of eq. (**) is:

$$\{k = 8 + 9\mu \qquad \lambda = 4 - 7\mu\} \quad \mu \in \mathbb{Z} \qquad\qquad (***)$$

Since k and λ must be positive integers, $\mu = 0$, (why?), and thus the allowed values of k and λ are, $\boldsymbol{k = 8}$ and $\boldsymbol{\lambda = 4}$, and then,

$$\begin{cases} x = 7k + 1 = 7 \cdot 8 + 1 = 57 \\ y = 9\lambda + 7 = 9 \cdot 4 + 7 = 43 \end{cases}$$

The sought for numbers are: $x = 57, y = 43$.

Example 2-2-3: Solve the Diophantine system:

$$\begin{cases} x + y = 3(z + w) \\ x + z = 3(y + w) \\ x + w = 3(y + z) \end{cases}$$

Solution: Solving the first equation for x, we find:

$$x = -y + 3z + 3w \qquad\qquad (*)$$

Substituting this expression of x, into the second and the third equation of the system, yields: $y = z$ and $y = w$. This means that $\boldsymbol{y = z = w}$, and then, from eq. (*) we find, $x = 5y$. Thus, the general solution of the system is obtained by setting $y = z = w = k$ and $x = 5k$, with $k \in \mathbb{Z}$, i.e.

$$\begin{cases} x = 5k \\ y = k \\ z = k \\ w = k \end{cases} \quad k = 0, \pm 1, \pm 2, \pm 3, \ldots$$

PROBLEMS

2-2-1) Solve the Diophantine system:

$$\begin{cases} 2x + y + z + w = 11 \\ x + 2y + z + w = 12 \\ x + y + 2z + w = 13 \end{cases}$$

(**Ans:** $x = -1 + k, y = k, z = 1 + k, w = 12 - 4k, \quad k \in \mathbb{Z}$).

2-2-2) Solve the Diophantine system:

$$\begin{cases} 2(x+y) = 3(z+w) \\ 3(x+z) = 4(y+w) \\ 4(x+w) = 5(y+z) \end{cases}$$

(Ans: $x = 229k, y = 149k, z = 131k, w = 121k, \quad k \in \mathbb{Z}$).

2-2-3) Solve the Diophantine system:

$$\begin{cases} x+y+z+w = 10 \\ 2x+3y+4z+w = 24 \\ 3x+2y+2z+w = 17 \end{cases}$$

(Ans: $x = k, y = 7-5k, z = 3k, w = 3+k, \quad k = 0, \pm 1, \pm 2, \pm 3, ...$).

2-2-4) Find the integer and positive solutions of the system:

$$\begin{cases} 11x - 5y = 7 \\ 3x + 11y - 5z = 7 \\ 19x - 3y + 4z - 3w = 34 \end{cases}$$

(Ans: General solution,

$$\begin{cases} x = -8 + 75k \\ y = -19 + 165k \\ z = -48 + 408k \\ w = -107 + 854k \end{cases} \quad k = 0, \pm 1, \pm 2, \pm 3, ...$$

Positive solutions are obtained for $k = 1, 2, 3, ...$).

CHAPTER 3: SECOND DEGREE DIOPHANTINE EQUATIONS

3-1) Some simple cases of second degree Diophantine equations

In this chapter we shall consider some rather simple cases of second degree Diophantine equations in two or three unknowns. The general form of a second degree equation in two unknowns, x and y is:

$$ax^2 + bxy + cy^2 + dx + ey + f = 0 \qquad (3-1-1)$$

The coefficients a, b, c, d, e, f are integers, and only integer solutions are allowed. A pair of integer numbers (x_0, y_0), with $x_0, y_0 \in \mathbb{Z}$ is a solution of the eq. (3-1-1) if it satisfies the equation, i.e. if

$$ax_0^2 + bx_0y_0 + cy_0^2 + dx_0 + ey_0 + f = 0 \qquad (3-1-2)$$

Solving the Diophantine equation means to find **all** its integer solutions.

Solving eq. (3-1-1) in its general form is a difficult subject. To simplify the problem we shall consider some special cases.

1) x^2 or y^2 is missing:

Let us assume that the term x^2 is missing. Equation (3-1-1) becomes:

$$bxy + cy^2 + dx + ey + f = 0 \qquad (3-1-3)$$

This is a first degree equation in x, which when solved for x yields:

$$x = \frac{-cy^2 - ey - f}{by + d} \qquad (3-1-4)$$

The numerator (in eq. (3-1-4)) is a second degree polynomial in y, while the denominator is a first degree polynomial in y. Performing the division, let $q(y) = ky + \lambda$ be the quotient of the division and r be the corresponding remainder, (note that since the divisor $by + d$ is a first degree polynomial in y, **the remainder r will be a constant number** (independent of y)). According to the equality of the algorithmic division, we have:

$$-cy^2 - ey - f = (by + d)(ky + \lambda) + r \qquad (3-1-5)$$

or, equivalently,

$$x = \frac{-cy^2 - ey - f}{by + d} = ky + \lambda + \frac{r}{by + d} \qquad (3-1-6)$$

If k, λ and r are not all integers, then they will be rational numbers, (fractions), and let D be their common denominator. Multiplying both sides of eq. (3-1-6) by D we find

$$Dx = Dky + D\lambda + \frac{Dr}{by + d}$$

or, if we call, $K = Dk$, $\Lambda = D\lambda$ and $R = Dr$, (all integers),

$$Dx = Ky + \Lambda + \frac{R}{by + d} \qquad (3-1-7)$$

a) Let us assume that $R \neq 0$. Then since Dx and $(Ky + \Lambda)$ are integers, the fraction $R/(by + d)$ must likewise be some integer, i.e. **the number $(by + d)$ must be a divisor of R**. If $R_1, R_2, \ldots, R_n$ are **all** the divisors of R, then setting

$$by + d = R_1, \qquad by + d = R_2, \qquad \ldots, \qquad by + d = R_n$$

we find **the integer values of y**, (the non integer values are rejected), and then, from eq. (3-1-7) we find **the associated integer values of x**, (the non integer values of x are rejected), (see Ex. 3-1-1).

b) If $R = 0$, i.e. $r = 0$, (since $R = Dr$), eq. (3-1-5) yields:

$$-cy^2 - ey - f = (by + d)(ky + \lambda) \overset{(3-1-4)}{\Longrightarrow}$$

$$x(by + d) = (by + d)(ky + \lambda) \Rightarrow (by + d)(x - ky - \lambda) = 0 \Rightarrow$$

$$\left\{ \begin{array}{c} by + d = 0 \\ or \\ x - ky - \lambda = 0 \end{array} \right\} \qquad (3-1-8)$$

The first equation in eq. (3-1-8) has integer solution **if and only if b divides d**, and in this case the solution is: $y = -d/b$ and $x = \boldsymbol{arbitrary\ integer}$. The second equation is a first order equation in x and y and can be solved as described in section 1-2, (see Ex. 3-1-2).

2) The equation can be expressed as: $(Ax + By + C)(Dx + Ey + F) = G$, **where** A, B, C, D, E, F, G **are all integers:**

In this case, the factors $(Ax + By + C)$ and $(Dx + Ey + F)$ must be factors of G, so if G can be factored **as the product of two other integers**, say $G = G_1 \cdot G_2$, then, from the original equation it follows that:

$$\left\{\begin{array}{c} Ax + By + C = G_1 \quad and \quad Dx + Ey + F = G_2 \\ or \\ Ax + By + C = G_2 \quad and \quad Dx + Ey + F = G_1 \end{array}\right\} \qquad (3-1-9)$$

From the first equation in (3-1-9) we find x and y, and **if both are integers they are accepted**, and in fact constitute one solution of the Diophantine equation $(Ax + By + C)(Dx + Ey + F) = G$. The same holds true with the second equation in eq. (3-1-9).

We repeat the foregoing analysis for **all possible factorizations of G in product of two integers**, (see Ex. 3-1-3).

Comment: The equation $ax^2 + bxy + cy^2 + dx + ey + f = 0$ can always be expressed in the form $(Ax + By + C)(Dx + Ey + F) = G$, if the quantity $b^2 - 4ac$ is equal to the square of some **integer** $m \neq 0$, i.e. if $b^2 - 4ac = m^2$.

Since in this case, **we may always find a number h, such that the polynomial $ax^2 + bxy + cy^2 + dx + ey + h$ can be expressed as a product of two first degree polynomials**. Indeed, we may express this polynomial as a quadratic polynomial in x:

$$ax^2 + (by + d)x + cy^2 + ey + h = a(x - \rho_1)(x - \rho_2) \qquad (3-1-10)$$

where ρ_1 and ρ_2 are the two roots of the quadratic polynomial in (3-1-10), given by the formula:

$$\rho_{1,2} = \frac{-(by + d) \pm \sqrt{\Delta}}{2a} \qquad (3-1-11)$$

In eq. (3-1-11), Δ is the discriminant of the quadratic equation, given by the formula

$$\Delta = (by + d)^2 - 4a(cy^2 + ey + h) \Rightarrow$$

$$\Delta = (b^2 - 4ac)y^2 + 2(bd - 2ae)y + d^2 - 4ah \qquad (3-1-12)$$

If $b^2 - 4ac = m^2$, the equation (3-1-12) is written as

$$\Delta = m^2 y^2 + 2(bd - 2ae)y + d^2 - 4ah \qquad (3-1-13)$$

Now, we may determine h so that the discriminant Δ in eq. (3-1-13) to be the square of some first degree polynomial $(ky + \lambda)$, i.e. $\Delta = (ky + \lambda)^2$. Indeed, the quadratic polynomial in y, in eq. (3-1-13), is a perfect square **provided that its discriminant is zero**, i.e. if

$$(bd - 2ae)^2 - m^2(d^2 - 4ah) = 0 \qquad (3-1-14)$$

From equation (3-1-14) **the number h is determined**, and thus

$$ax^2 + bxy + cy^2 + dx + ey + f = 0 \Longrightarrow$$

$$ax^2 + bxy + cy^2 + dx + ey + h = h - f \overset{\substack{(3-1-10)\\(3-1-11)}}{\Longrightarrow}$$

$$a\left\{x - \frac{-(by + d) + (ky + \lambda)}{2a}\right\}\left\{x - \frac{-(by + d) - (ky + \lambda)}{2a}\right\} = h - f$$

which is of the form $(Ax + By + C)(Dx + Ey + F) = G$, (see Ex. 3-1-4).

3) The case of a finite number of trials:

When the equation $ax^2 + bxy + cy^2 + dx + ey + f = 0$ is solved for x, we find:

$$x = \frac{-(by + d) \pm \sqrt{(by + d)^2 - 4 \cdot a \cdot (cy^2 + ey + f)}}{2a} \qquad (3-1-15)$$

Since we seek for integer solutions (x, y), **the radicand must be a positive quantity, and in addition, must be a perfect square**. So, we start by finding the values of y that make $(by + d)^2 - 4 \cdot a \cdot (cy^2 + ey + f) \geq 0$. This is a quadratic trinomial in y. **If the coefficient of y^2 is negative**, then, the roots of this trinomial, ρ_1 and ρ_2, must be real and y must lie between the two roots, i.e. $\rho_1 \leq y \leq \rho_2$. We find the integer values of y that lie between the roots. From these values, we choose the ones that make the radicand **a perfect square**, and then, among them, we choose the ones that yield integer values of

x, (in eq. (3-1-15)). In case the coefficient of y^2 in the radicand is positive, we may solve the equation for y, check whether the coefficient of x^2 in the radicand is negative, and repeat the aforementioned procedure, (see Ex 3-1-5).

4) The method of substitution:

Sometimes, the form of the equation is such that, by means **a proper substitution**, the original equation is reduced to a simpler one which is solved easily. For example, the equation $y^2 - 2xy + x^2 + y - x - 2 = 0$ may be written as $(y - x)^2 + (y - x) - 2 = 0$, and if we call $z = y - x$, the equation is reduced to $z^2 + z - 2 = 0$ which gives, $z = 1$ or $z = -2$, i.e. $y - x = 1$ or $y - x = -2$, which are solved easily, (see Ex. 3-1-6).

As another example, let us consider the equation:
$y^2 + 2xy + x^2 + y - x - 4 = 0$, which may be written as $(y + x)^2 + (y - x) - 4 = 0$, or, if we make the substitution $z = y + x$ and $w = y - x$, the equation is reduced to $z^2 + w - 4 = 0$, (see Ex. 3-1-7).

Example 3-1-1: Solve the Diophantine equation
$$3xy + 2y^2 + 5x - 7y = 11$$

Solution: This is a second degree Diophantine equation, where the term x^2 is missing. Solving for x we find:

$$x = \frac{-2y^2 + 7y + 11}{3y + 5} \qquad (*)$$

Performing the division in eq. (*) we find:

$$x = -\frac{2}{3}y + \frac{31}{9} - \frac{\frac{56}{9}}{3y + 5}$$

or, if we multiply through by 9,

$$9x = -6y + 31 - \frac{56}{3y + 5} \qquad (**)$$

Since $9x$ and $(-6y + 31)$ are integers, the fraction $56/(3y + 5)$ must likewise be some integer, which means that $(3y + 5)$ must divide exactly 56.

The divisors of 56 are: $\pm 1, \pm 2, \pm 4, \pm 7, \pm 8, \pm 14, \pm 28, \pm 56$, so these are the possible values of $(3y + 5)$.

If $3y + 5 = 1$, $y = -4/3$, (rejected, not integer).

If $3y + 5 = -1$, $y = -2$, (**accepted, integer**).

If $3y + 5 = 2$, $y = -1$, (**accepted, integer**).

If $3y + 5 = -2$, $y = -7/3$, (rejected, not integer).

If $3y + 5 = 4$, $y = -1/3$, (rejected, not integer).

If $3y + 5 = -4$, $y = -3$, (**accepted, integer**).

If $3y + 5 = 7$, $y = 2/3$, (rejected, not integer).

If $3y + 5 = -7$, $y = -4$, (**accepted, integer**).

If $3y + 5 = 8$, $y = 1$, (**accepted, integer**).

If $3y + 5 = -8$, $y = -13/3$, (rejected, not integer).

If $3y + 5 = 14$, $y = 3$, (**accepted, integer**).

If $3y + 5 = -14$, $y = -19/3$, (rejected, not integer).

If $3y + 5 = 28$, $y = -23/3$, (rejected, not integer).

If $3y + 5 = -28$, $y = -11$, (**accepted, integer**).

If $3y + 5 = 56$, $y = 17$, (**accepted, integer**).

If $3y + 5 = -56$, $y = -61/3$, (rejected, not integer).

Thus, the allowed values of y are: $-2, -1, -3, -4, 1, 3, -11, 17$.

The integer values of x, (if there are any), are determined from eq. (**). When $y = -2$, eq. (**) yields:

$$9x = -6 \cdot (-2) + 31 + 56 = 99 \implies x = 11$$

Thus, one solution of the equation is $(x = 11, y = -2)$.

Working similarly, with **all the accepted values of y**, we find all the integer solutions of our equation, (details in Pr. 3-1-1):

$$(x, y) = (11, -2), (1, -1), (7, -3), (7, -4), (2,1), (3,3), (11, -11), (-8,17)$$

Example 3-1-2: Solve the Diophantine equation

$$10y^2 - 2xy + 6x - 24y - 18 = 0$$

Solution: Since the term x^2 is missing, we solve for x:

$$x = \frac{10y^2 - 24y - 18}{2y - 6} \qquad (*)$$

and performing the division we find, $x = 5y + 3$, (perfect division with remainder $r = 0$). Thus,

$$x = \frac{10y^2 - 24y - 18}{2y - 6} = 5y + 3 \Longrightarrow$$

$$x(2y - 6) = (2y - 6)(5y + 3) \Longrightarrow (2y - 6)(x - 5y - 3) = 0 \Longrightarrow$$

$$\{2y - 6 = 0 \quad \textbf{or} \quad x - 5y - 3 = 0\} \qquad (**)$$

The solution of the first equation is: $x = k$ (**arbitary integer**), $y = 3$.

From the second equation, $x = 5y + 3$, and its general solution is:

$$x = 5\lambda + 3, \qquad y = \lambda \qquad (\textbf{arbitrary integer})$$

Thus, the solutions of the given equation are:

$$\begin{cases} x = k \text{ (arbitary integer)}, & y = 3 \\ x = 5\lambda + 3, & y = \lambda \quad (\textit{arbitrary integer}) \end{cases}$$

Example 3-1-3: Find the integer solutions of the equation

$$(3x - 2y + 1)(2x + 5y - 7) = 10 \qquad (*)$$

Solution: The number 10 can be factored in the following ways:

$$1 \cdot 10, (-1) \cdot (-10), 2 \cdot 5, (-2) \cdot (-5), 5 \cdot 2, (-5) \cdot (-2), 10 \cdot 1, (-10) \cdot (-1)$$

1) With $10 = 1 \cdot 10$, the given equation is satisfied if:

$$\{3x - 2y + 1 = 1 \quad \textbf{\textit{and}} \quad 2x + 5y - 7 = 10\}, \quad i.e.$$

$$\{3x - 2y = 0 \quad \textbf{\textit{and}} \quad 2x + 5y = 17\}$$

This is a system of two linear equations in two unknowns. The reader is supposed to know how to solve such a system. Solving the system, (for example, solving the first equation for y and substituting in the second), we find $(x = 34/19, y = 51/19)$. This solution is rejected, since x and y, are not integers.

2) With $10 = (-1) \cdot (-10)$, eq. (*) is satisfied if:

$$\{3x - 2y + 1 = -1 \quad \textbf{\textit{and}} \quad 2x + 5y - 7 = -10\}, \quad i.e.$$

$$\{3x - 2y = -2 \quad \textbf{\textit{and}} \quad 2x + 5y = -3\}$$

The solution is $(x = -16/19, y = -5/19)$, rejected (not integers).

Working similarly, with **all the other factorizations of 10**, we find that only the factorization $10 = 5 \cdot 2$ leads to integer solutions, and all the others lead to not integer solutions, and therefore are rejected. For $10 = 5 \cdot 2$, eq. (*) leads to

$$\{3x - 2y + 1 = 5 \quad \textbf{\textit{and}} \quad 2x + 5y - 7 = 2\}, \quad or$$

$$\{3x - 2y = 4 \quad \textbf{\textit{and}} \quad 2x + 5y = 9\}$$

The solution of this system is $(x = 2, y = 1)$, and this is the only solution of the Diophantine equation.

Example 3-1-4: Solve the Diophantine equation

$$2x^2 - xy - y^2 - 5x + 8y - 25 = 0$$

Solution: In this problem, $a = 2, b = -1, c = -1$, and the quantity $b^2 - 4ac = (-1)^2 - 4 \cdot 2 \cdot (-1) = 9 = 3^2$. Therefore, we can determine a number h, such that the left side of the equation

$$2x^2 - xy - y^2 - 5x + 8y - 25 + h = h \qquad (*)$$

can be written **as the product of two linear equations**. The left side of equation (*) is written as:

$$2x^2 - (y+5)x - y^2 + 8y + h - 25 \qquad (**)$$

If ρ_1, ρ_2 are the two roots of the quadratic trinomial in eq. (**), we may write:

$$2x^2 - (y+5)x - y^2 + 8y + h - 25 = 2(x - \rho_1)(x - \rho_2) \qquad (***)$$

The two roots, ρ_1 and ρ_2, of the trinomial in (**) are given by the formula:

$$\rho_{1,2} = \frac{-\{-(y+5)\} \pm \sqrt{\Delta}}{2 \cdot 2} = \frac{y + 5 \pm \sqrt{\Delta}}{4} \qquad (****)$$

where the discriminant $\Delta = (y+5)^2 - 4 \cdot 2 \cdot (-y^2 + 8y + h - 25)$, which after some simplifications takes the form

$$\Delta = 9y^2 - 54y + 225 - 8h \qquad (*****)$$

We want to determine h, so that the discriminant Δ to be the square of some first degree polynomial $(ky + \lambda)$, (i.e. $\boldsymbol{\Delta = (ky + \lambda)^2}$). The necessary condition for this is the discriminant of the trinomial $(9y^2 - 54y + 225 - 8h)$ to be zero, i.e.

$$(-54)^2 - 4 \cdot 9 \cdot (225 - h) = 0 \Rightarrow h = 18$$

With this value for h, eq. (*****) becomes:

$$\Delta = 9y^2 - 54y + 81 = (3y - 9)^2$$

Then, from eq. (****) it follows

$$\rho_{1,2} = \frac{y + 5 \pm \sqrt{\Delta}}{4} = \frac{y + 5 \pm (3y - 9)}{4} \Rightarrow \left\{ \begin{array}{l} \rho_1 = y - 1 \\ \rho_2 = (-y + 7)/2 \end{array} \right\}$$

Having found ρ_1 and ρ_2, eq. (***) becomes, (since $h = 18$):

$$2x^2 - (y+5)x - y^2 + 8y + 18 - 25 = 2(x - y + 1)\left(x - \frac{-y + 7}{2}\right) \Rightarrow$$

$$2x^2 - (y+5)x - y^2 + 8y + 18 - 25 = (x - y + 1)(2x + y - 7)$$

and then, from eq. (*),

$$(x - y + 1)(2x + y - 7) = 18 \qquad (\text{******})$$

The number 18 can be factored in the following ways, (12 ways):

$$1 \cdot 18, (-1) \cdot (-18), 2 \cdot 9, (-2) \cdot (-9), 3 \cdot 6, (-3) \cdot (-6), 6 \cdot 3, (-6) \cdot (-3),$$
$$9 \cdot 2, (-9) \cdot (-2), 18 \cdot 1, (-18) \cdot (-1)$$

So, eq. (******) is split into 12 equations. For example, if we write $18 = 3 \cdot 6$, eq. (******) implies,

$$x - y + 1 = 3 \quad \textbf{and} \quad 2x + y - 7 = 6$$

Solving this system we find: $x = 5, y = 3$, and this is an accepted solution (integer solution). Had we chosen another factorization of 18, say $18 = 1 \cdot 18$, then, from eq. (******) we would have:

$$x - y + 1 = 1 \quad and \quad 2x + y - 7 = 18$$

The solution of this system is $x = y = 25/3$, which is rejected, since it is not an integer solution. For the solution of eq. (******) for all possible factorizations of 18, see Pr. 3-1-2.

Example 3-1-5: Solve the Diophantine equation

$$3x^2 + 4xy + 2y^2 - 6x - 4y + 2 = 0$$

Solution: We may write the equation as

$$2y^2 + 4(x - 1)y + 3x^2 - 6x + 2 = 0 \qquad (*)$$

Solving this equation for y yields:

$$y = \frac{-4(x - 1) \pm \sqrt{16(x - 1)^2 - 4 \cdot 2 \cdot (3x^2 - 6x + 2)}}{2 \cdot 2} \Longrightarrow$$

$$y = \frac{-2(x - 1) \pm \sqrt{2(-x^2 + 2x)}}{2} \qquad (**)$$

Since y must be real, the discriminant $2(-x^2 + 2x)$ must be a positive number, or zero, i.e.

$$2(-x^2 + 2x) \geq 0 \Leftrightarrow x^2 - 2x \leq 0 \Leftrightarrow x(x-2) \leq 0 \Leftrightarrow 0 \leq x \leq 2$$

The **integer** values of x in this interval, are: $x = 0, 1, 2$.

For $x = 0$, we find from eq. (**): $y = 1$. Thus $(x = 0, y = 1)$ is one solution of the equation.

For $x = 1$, we find from eq. (**): $y = \pm \sqrt{2}/2$, (rejected, not an integer).

For $x = 2$, we find from eq. (**): $y = -1$. Thus $(x = 2, y = -1)$ is another solution of the equation.

Summarizing, the given equation has two integer solutions: $(x = 0, y = 1)$ and $(x = 2, y = -1)$.

Example 3-1-6: Solve the Diophantine equation

$$y^2 - 2xy + x^2 + y - x - 2 = 0$$

Solution: The given equation is written as $(y - x)^2 + (y - x) - 2 = 0$, or, if we set $z = y - x$, the equation becomes $z^2 + z - 2 = 0$, whose solutions are easily found to be, $z = 1$, or, $z = -2$.

a) When $z = 1$, $y - x = 1$, or $y = 1 + x$. The general solution of this equation is $\{x = k, \quad y = 1 + k\}, \quad k = 0, \pm 1, \pm 2, \pm 3, \dots$

b) When $z = -2$, $y - x = -2$, or, $y = -2 + x$, and the general solution of this equation is $\{x = \lambda, \quad y = -2 + \lambda\}, \quad \lambda = 0, \pm 1, \pm 2, \pm 3, \dots$

Example 3-1-7: Solve the Diophantine equation

$$y^2 + 2xy + x^2 + y - x - 4 = 0$$

Solution: The given equation is written as $(y + x)^2 + (y - x) - 4 = 0$, or, if we set $z = y + x$ and $w = y - x$, the equation becomes $z^2 + w - 4 = 0$, or $w = 4 - z^2$. The integer solution of this equation is $\{z = k, \quad w = 4 - k^2\}, \quad k \in \mathbb{Z}$, i.e.

$$\left\{ \begin{array}{l} y + x = k \\ y - x = 4 - k^2 \end{array} \right\} \Rightarrow \left\{ y = -2 + \frac{k(k+1)}{2}, \quad x = 2 - \frac{k(k-1)}{2} \right\}, k \in \mathbb{Z}$$

Note that for any $k \in \mathbb{Z}$, x and y are integers, since the product of two consecutive integers is a multiple of 2, i.e. $k(k \pm 1)/2$ is integer.

Example 3-1-8: Solve the Diophantine equation $x^2 - y^2 = 3$.

Solution: The given equation is written as $(x - y)(x + y) = 3$, and since $3 = 1 \cdot 3$, or $(-1) \cdot (-3)$, or $3 \cdot 1$, or $(-3) \cdot (-1)$, it follows:

$$\left. \begin{cases} x - y = 1 \\ x + y = 3 \end{cases} \right\} \ or \ \left. \begin{cases} x - y = -1 \\ x + y = -3 \end{cases} \right\} \ or \ \left. \begin{cases} x - y = 3 \\ x + y = 1 \end{cases} \right\} \ or \ \left. \begin{cases} x - y = -3 \\ x + y = -1 \end{cases} \right\}$$

Solving the systems we find the integer solutions: $(x = 2, y = 1)$, $(x = -2, y = -1)$, $(x = 2, y = -1)$, $(x = -2, y = 1)$.

Example 3-1-9: Solve the Diophantine equation: $(x + 1)(y + 2) = 2xy$.

Solution: $(x + 1)(y + 2) = 2xy$, or, $xy + y + 2x + 2 = 2xy$, or

$2xy - xy - y - 2x = 2$, or $xy - y - 2x = 2$, or, $x(y - 2) - y + 2 = 4$, or

$x(y - 2) - (y - 2) = 4$, or, $(x - 1)(y - 2) = 4$. This equation is split to the following equations:

$$\left. \begin{cases} x - 1 = 1 \\ y - 2 = 4 \end{cases} \right\} \ or \ \left. \begin{cases} x - 1 = -1 \\ y - 2 = -4 \end{cases} \right\} \ or \ \left. \begin{cases} x - 1 = 2 \\ y - 2 = 2 \end{cases} \right\} \ or \ \left. \begin{cases} x - 1 = -2 \\ y - 2 = -2 \end{cases} \right\} \ or$$

$$\left. \begin{cases} x - 1 = 4 \\ y - 2 = 1 \end{cases} \right\} \ or \ \left. \begin{cases} x - 1 = -4 \\ y - 2 = -1 \end{cases} \right\}$$

Solving these equations we find all the integer solutions of the original equation:

$$(x, y) = (2,6), (0, -2), (3,4), (-1,0), (5,3), (-3,1)$$

Example 3-1-10: Find the non negative integer solutions of the Diophantine equation: $(x + y)^2 = 10x + y$.

Solution: The equation is written as

$$x^2 + 2xy + y^2 - 10x - y = 0, \ or, \ x^2 + 2(y - 5)x + y^2 - y = 0$$

Solving for x we find:

$$x = \frac{-2(y-5) \pm \sqrt{4(y-5)^2 - 4 \cdot 1 \cdot (y^2 - y)}}{2} \Rightarrow$$

$$x = -(y-5) \pm \sqrt{25 - 9y} \qquad (*)$$

First of all, the radicand must be greater than or equal to zero, i.e. $25 - 9y \geq 0$, or $y \leq 25/9$ and since $y \geq 0$, (non negative solutions), $0 \leq y \leq 25/9$. Thus, the allowed values of y are: $\mathbf{y = 0, 1, 2}$. Among these values we must choose those that make $\sqrt{25 - 9y}$ to be an integer. These values are $y = 0$ and $y = 1$, (the value $y = 2$ is rejected, since $\sqrt{25 - 9 \cdot 2} = \sqrt{7}$ would yield not integer x).

a) For $y = 0$, the corresponding values of x are, (from eq. (*)): $x = 10, \ 0$.

b) For $y = 1$, the corresponding values of x are, (from eq. (*)): $x = 8, \ 0$.

Thus, the non negative solutions of the given equation are:

$$(x, y) = (10,0), (0,0), (8,1), (0,1)$$

3-1-11) Find all the positive integers which satisfy the following relations simultaneously: $\{5x - 8y = 12, \quad xy - 4y < 400\}$.

Solution: Since $x = 4, y = 1$ is one solution of the first equation, its general solution is:

$$x = 4 - 8k, \qquad y = 1 - 5k, \quad k \in \mathbb{Z} \qquad (*)$$

Since we seek positive solutions, $x > 0$ and $y > 0$, i.e. $(4 - 8k) > 0$ and $(1 - 5k) > 0$, i.e. $k < 1/2$ and $k < 1/5$. The integer k which satisfy both inequalities **simultaneously** are: $\mathbf{k = 0, -1, -2, -3, ...}$

The solutions we seek must also satisfy the inequality, $(xy - 4y) < 400$, which by virtue of eq. (*) yields:

$$(4 - 8k)(1 - 5k) - 4(1 - 5k) < 400 \Longleftrightarrow 40k^2 - 8k - 400 < 0 \Longleftrightarrow$$

$$5k^2 - k - 50 < 0 \qquad (**)$$

The inequality in eq. (**) is satisfied for $\rho_1 < k < \rho_2$, where $\rho_{1,2}$ are the two roots of the trinomial $5k^2 - k - 50$, given by the formula

$$\rho_{1,2} = \frac{-(-1) \pm \sqrt{(-1)^2 - 4 \cdot 5 \cdot (-50)}}{2 \cdot 5} = \frac{1 \pm \sqrt{1001}}{10} \Longrightarrow \begin{cases} \rho_1 \cong 3.26 \\ \rho_2 \cong -3.06 \end{cases}$$

Thus, $-3.06 < k < 3.26$, and since k must be either zero or a negative integer, as found previously, the allowed values of k are: $k = 0, -1, -2, -3$. Therefore, there are four pairs of integers (x, y), as shown in the following table:

	$x = (4 - 8k)$	$y = (1 - 5k)$
$k = 0$	4	1
$k = -1$	12	6
$k = -2$	20	11
$k = -3$	28	16

Example 3-1-12: Solve the Diophantine equation $6y = 5x^2 + 4x + 6$.

Solution: We may write this equation as follows:

$$6y = 6x^2 + 6x + 6 - x^2 - 2x = 6(x^2 + x + 1) - x(x + 2) \Longrightarrow$$

$$y = x^2 + x + 1 - \frac{x(x + 2)}{2 \cdot 3} \qquad (*)$$

This equation will be satisfied by integers x and y, **provided that** $x(x + 2)/(2 \cdot 3)$ **is an integer**, and since 2 and 3 are prime numbers, this means that either 2 divides x **and** 3 divides $(x + 2)$, or, 2 divides $(x + 2)$ **and** 3 divides x.

Case 1: $2 \, / \, x$ and $3/(x + 2)$. This means that $x = 2k$ **and** $x + 2 = 3\lambda$, with $k, \lambda \in \mathbb{Z}$. However, the integers k and λ cannot be unrelated, since from $x = 2k$ and $x + 2 = 3\lambda$, or, $x = 3\lambda - 2$ it follows that

$$2k = 3\lambda - 2, \quad or, \quad 2k - 3\lambda = -2 \qquad (**)$$

Equation (**) is a Diophantine equation in k and λ. One solution of this equation is $k = 2, \lambda = 2$, and therefore its general solution is

$$\{k = 2 - 3t \qquad \lambda = 2 - 2t\}, \quad t = 0, \pm1, \pm2, \pm3, \dots \qquad (***)$$

Then, $x = 2k = 4 - 6t$, and substituting this expression of x in eq. (*), we find $y = 30t^2 - 44t + 17$. Thus one solution of the given equation is:

$$\{x = 4 - 6t \qquad y = 30t^2 - 44t + 17\}, \quad t = 0, \pm1, \pm2, \pm3, \dots \quad (****)$$

For example, for $t = 0$, we find one solution $(x, y) = (4,17)$, for $t = 1$ we find another solution $(x, y) = (-2,3)$, etc.

Case 2: $2 \,/\, (x + 2)$ and $3 \,/\, x$. Working as in case 1, we must have, $x + 2 = 2k, x = 3\lambda, \ k, \lambda \in \mathbb{Z}$, and therefore, $2k - 2 = 3\lambda$, i.e. **$2k - 3\lambda = 2$**. This is a Diophantine equation in k and λ. One solution of this equation is $(k = 4, \lambda = 2)$, and thus the general solution is, **$k = 4 - 3w, \lambda = 2 - 2w$**, with $w \in \mathbb{Z}$. It follows that

$$\{x = 6 - 6w \qquad y = 30w^2 - 64w + 35\}, w = 0, \pm1, \pm2, \dots \quad (*****)$$

This is another set of solutions. For example, for $w = 0$ we find one solution $(x, y) = (6,35)$, for $w = -1$ we find another solution $(x, y) = (12,129)$, etc.

Summarizing, the given equation has two solution sets, given by formulas (****) and (*****).

PROBLEMS

3-1-1) Verify the integer solutions found in Ex. 3-1-1.

3-1-2) Solve eq. (******) in Ex. 3-1-4, for all possible factorizations of the number 18, and subsequently, find all the integer solutions of the given Diophantine equation.

3-1-3) Solve the Diophantine equation: $(x + 5)(y + 6) = 3xy$.

(**Ans:** $(x, y) = (25,4), (3,48), (10,6), (4,18), (7,8), (5,12)$

Hint: See Ex. 3-1-9.

3-1-4) Solve the Diophantine equation: $y^2 - 9x^2 = 7$.

Hint: $(y + 3x)(y - 3x) = 7$. The possible factorizations of 7 are: $1 \cdot 7, (-1) \cdot (-7), 7 \cdot 1, (-7) \cdot (-1)$.

3-1-5) Solve the Diophantine equation: $x^2 + 2xy + y^2 + y - x = 10$.

(Ans: $x = -5 + k(k+1)/2$, $y = 5 - k(k-1)/2$, $k \in \mathbb{Z}$).

Hint: Set $z = y + x$, $w = y - x$, (See Ex. 3-1-7).

3-1-6) Find the integer values of x and y which satisfy the equation:

$$2y^2 - 2x + 3y - 1 = 0$$

(Ans: $x = 4k^2 + 15k + 13, y = 3 + 2k$, $k \in \mathbb{Z}$).

Hint: $2x = 2y^2 + 3y - 1 = 2y^2 + 2y + 2 + (y - 3)$, or,
$x = y^2 + y + 1 + (y - 3)/2$, and this equation has integer solutions provided that $(y - 2)/3 = k \in \mathbb{Z}$, i.e. $y = 3k + 2$, etc.

3-1-7) Find all the integer solutions of the eq: $x^2 - 3xy + x + y + 1 = 0$.

(Ans: $(x, y) = (0, -1), (-4, -1)$).

Hint: The term y^2 is missing, (see Ex. 3-1-1).

3-1-8) Find all the integer solutions of the equation:

$$(3xy + y - 8)(2xy + 3x - 10) = 0$$

Hint: The given eq. implies, $(3xy + y - 8) = 0$, or, $(2xy + 3x - 10) = 0$. The first eq. is written as $(3x + 1)y = 8$. The divisors of 8 are: $\pm 1, \pm 2, \pm 4, \pm 8$, and therefore, $3x + 1 = 1$ and $y = 8$, or, $3x + 1 = -1$ and $y = -8$, etc. The same for the second equation, (see Ex. 3-1-3).

3-1-9) Find all the integer solutions of the equation:

$$x(3 - |y|) + y(3 - |x|) + |xy| = 6$$

(Ans: $(x, y) = (6,4), (0,2), (4,6), (2,0), (-4, -2), (-2, -4)$).

Hint: Consider the four cases: $x \geq 0, y \geq 0, x \geq 0, y < 0, x < 0, y \geq 0$ and $x < 0, y < 0$. In the first case, $|x| = x, |y| = y$, in the second case $|x| = x, |y| = -y$, etc.

3-1-10) Find the positive and integer solutions of the Diophantine equation: $3xy - x^2 - 10y + 4x + 12 = 0$.

(Ans: $(x, y) = (3,15), (2,4), (14,4), (46,15))$.

Hint: The term y^2 is missing, see Ex. 3-1-1.

3-2) The Diophantine equation $x^2 + ay^2 = z^2$, with $a \in \mathbb{Z}$

We may always assume that $a > 0$, since:

If $a = 0$, the equation becomes $x^2 = z^2$, i.e. $x = \pm z$, (integer), and y being an arbitrary integer, while if $a < 0$, the equation becomes, $z^2 + (-a)y^2 = x^2$, with $-a > 0$, which is of the same form with the original equation, with x and z interchanged. Therefore, we shall study the equation $x^2 + ay^2 = z^2$ with $a > 0$, integer).

a) One obvious, integer solution, is $x = 0, y = 0, z = 0$. Also, if $y = 0$, the equation reduces to $x^2 = z^2$, i.e. $x = \pm z$. We thus obtain another obvious solution, $(x = k, y = 0, z = k)$, or, $(x = -k, y = 0, z = k)$, where $k \in \mathbb{Z}$.

Let us now assume that $y \neq 0$. Dividing both sides of the equation by y^2, we find:

$$\left(\frac{x}{y}\right)^2 + a = \left(\frac{z}{y}\right)^2 \qquad (3-2-1)$$

Since $a > 0$, we may set:

$$\frac{z}{y} = \frac{x}{y} + \frac{m}{n}, \qquad where \ (m,n) = 1 \ and \ m, n \in \mathbb{N} \qquad (3-2-2)$$

i.e. **the m and n are relatively prime integers**, (i.e. the fraction m/n is expressed in its lowest terms). Substituting this expression of z/y into equation (3-2-1), we find:

$$a = \frac{m^2}{n^2} + \frac{2mx}{ny} \Longleftrightarrow \frac{2mx}{ny} = a - \frac{m^2}{n^2} = \frac{an^2 - m^2}{n^2} \Longleftrightarrow$$

$$\frac{x}{y} = \frac{an^2 - m^2}{2mn} \qquad\qquad (3-2-3)$$

Equation (3-2-3) is satisfied if

$$x = (an^2 - m^2)k, \quad y = 2mn\,k, \qquad k \in \mathbb{Z} \qquad\qquad (3-2-4)$$

Solving eq. (3-2-2) for z, and substituting the expressions of x and y, (in terms of n and m), we find, $z = (an^2 + m^2)k$.

Summarizing, the integer solutions of $x^2 + ay^2 = z^2$ are given by the formulas:

$$\{x = (an^2 - m^2)k, \quad y = 2mn\,k, \quad z = (an^2 + m^2)k\}, \quad k, m, n \in \mathbb{Z}$$

$$(3-2-5)$$

b) The Pythagorean equation $x^2 + y^2 = z^2$.

This equation is known as **the Pythagorean equation**, and is a special case of the equation $x^2 + ay^2 = z^2$, when $a = 1$. Therefore, the solution of the Pythagorean equation is given by the formulas (3-2-5) with $a = 1$, i.e.

$$\{x = (n^2 - m^2)k, \quad y = 2mnk, \quad z = (n^2 + m^2)k\}, \quad k, m, n \in \mathbb{Z}$$

$$(3-2-6)$$

The problem of finding all the right triangles, whose sides are integer numbers, (**Pythagorean triangles**), leads to finding the integer solutions of the equation $x^2 + y^2 = z^2$, (see Ex. 3-2-2).

Every solution of the Pythagorean equation is known as "**a Pythagorean triple**". Special types of Pythagorean triples were known to the ancient Greeks. For example, some special, integer solutions of the equation $x^2 + y^2 = z^2$, are the following:

$$x = k, \quad y = \frac{k^2 - 1}{2}, \quad z = \frac{k^2 + 1}{2}, \quad k \text{ is an } \textbf{odd } integer - \{1\}$$

$$x = 2k, \quad y = k^2 - 1, \quad z = k^2 + 1, \quad k \in \mathbb{N} - \{1\}$$

$$x = 2k + 1, \qquad y = \frac{(2k+1)^2 - 1}{2}, \qquad z = \frac{(2k+1)^2 - 1}{2} + 1, \quad k \in \mathbb{N}$$

An interesting property is that **the product of the integers of any Pythagorean triple is divisible by 30**, (see Ex. 3-2-4).

Comment: Fermat's last theorem is a famous conjecture made by Fermat, in 1637, according to which, no three positive integers x, y and z satisfy the equation $x^n + y^n = z^n$ for integer values of n greater than 2. In other words, for integer $n > 2$, the equation $x^n + y^n = z^n$ does **not** have integer solutions. Even though Fermat claimed that he had found a really beautiful solution, but the margin of his notebook was too short to contain the proof, his proof was never found. Not only this, but since then, Fermat's conjecture resisted proof, until 1994, when the mathematician **Andrew Wiles** finally proved that Fermat's conjecture is indeed true.

Example 3-2-1: Solve the Diophantine equation $x^2 + 3y^2 = z^2$.

Solution: The solution of the equation is given by formula (3-2-5), with $a = 3$:

$$\{x = (3n^2 - m^2)k, \quad y = 2mn\,k, \quad z = (3n^2 + m^2)k\}, \quad k, m, n \in \mathbb{Z}$$

For example, for $k = 1, n = 3, m = 2$, we find:$(x, y, z) = (23,12,31)$, for $k = 2, n = 2, m = 1$, we find another Pythagorean triad, $(x, y, z) = (22,8,26)$, etc.

Example 3-2-2: Find all the right triangles whose sides are integer numbers.

Solution: If x and y are the two perpendicular sides of a right triangle and z is the hypotenuse, then, $z^2 = x^2 + y^2$, and from eq. (3-2-6),

$$\{x = (n^2 - m^2)k, \quad y = 2mn\,k, \quad z = (n^2 + m^2)k\}, \quad k, n, m \in \mathbb{N},$$

and $n > m$, (since the side $x > 0$). For example, for $k = 1, n = 2, m = 1$, we find $(x, y, z) = (3,4,5)$, which is most familiar triad of Pythagorean numbers, $(3^2 + 4^2 = 5^2)$.

Example 3-2-3: Show that the radius of the inscribed circle in a Pythagorean triangle is an integer number.

Solution

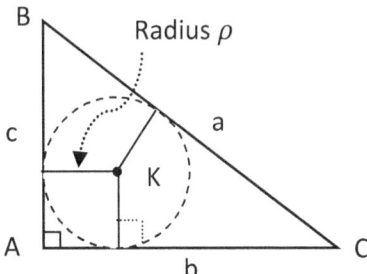

Fig. 3-1: Circle inscribed in a Pythagorean triangle.

Let ABC be a right triangle ($\hat{A} = 90°$). If E is the area of the triangle, and ρ is the radius of the inscribed circle, then, as we know from Geometry:

$$E = \tau\rho, \quad where \quad 2\tau = a + b + c \,(= perimeter \; of \; the \; trianle) \quad (*)$$

However, for a right triangle, the area is $E = bc/2$, and thus

$$\frac{bc}{2} = \tau\rho = \frac{(a + b + c)\rho}{2} \Longrightarrow \rho = \frac{bc}{a + b + c} \quad (**)$$

Since a, b, c form a Pythagorean triad, ($b^2 + c^2 = a^2$), we have, (from eq. (3-2-6)):

$$\rho = \frac{(n^2 - m^2)(2mn)k^2}{\{(n^2 - m^2) + 2mn + (n^2 + m^2)\}k} = \frac{(n^2 - m^2)(2mn)}{2n^2 + 2nm}k \Longrightarrow$$

$$\rho = \frac{2nm(n + m)(n - m)}{2n(n + m)}k = m(n - m)k \quad (\textbf{\textit{integer}})$$

(since n, m, k are integers, with $n - m > 0$).

Example 3-2-4: Show that the product of the integers of any Pythagorean triple is divisible by 30.

Solution: Let x, y, z be a Pythagorean triple. Then, by virtue of the formulas in eq. (3-2-6), we have:

$$xyz = (n^2 - m^2)(2nm)(n^2 + m^2)k^3, \quad n, m, k \in \mathbb{Z} \qquad (*)$$

If we show that the product xyz is divisible by 2, **and** by 3, **and** by 5, then the product will be divisible by the product $2 \cdot 3 \cdot 5 = 30$, since $2, 3, 5$ are prime numbers. **The product is obviously divisible by 2**, (due to the term $2mn$).

Let us show that the product is divisible by 3. If we divide the integers n and m by 3, we find (from the equality of the algorithmic division):

$$\left\{ \begin{array}{l} n = 3a, \quad or \quad n = 3a + 1, \quad or \quad n = 3a + 2 \\ m = 3b, \quad or \quad m = 3b + 1, \quad or \quad m = 3b + 2 \end{array} \right\} \quad a, b \ integers \qquad (**)$$

If $n = 3a$, **or** $m = 3b$, then the product xyz, (in eq. (*)), is divisible by 3.

If $n = 3a + 1$ **and** $m = 3b + 1$, then $(n^2 - m^2) = (n - m)(n + m)$, (in equation (*)), is divisible by 3, since $(n - m) = 3(a - b)$ is divisible by 3, and therefore the whole product xyz is divisible by 3. If $n = 3a + 1$ **and** $m = 3b + 2$, then, again the factor $(n^2 - m^2) = (n - m)(n + m)$ is divisible by 3, since $(n + m) = 3(a + b) + 3 = 3(a + b + 1)$, and therefore the product xyz is divisible by 3. Similarly, we show that if $n = 3a + 2$ **and** $m = 3b + 1$, the product xyz is divisible by 3. Thus, **the product xyz is always divisible by 3.**

Let us now show that the product is divisible by 5. If the integers n and m are divided by 5, then, from the equality of the algorithmic division, we have:

$$\left\{ \begin{array}{lllll} n = 5a & n = 5a + 1 & n = 5a + 2 & n = 5a + 3 & n = 5a + 4 \\ m = 5b & m = 5b + 1 & m = 5b + 2 & m = 5b + 3 & m = 5b + 4 \end{array} \right\} \quad (***)$$

where a, b are integers.

Working similarly, we may show easily that for any n and m the product xyz is divisible by 5. For example, if $n = 5a + 3$ and $m = 5b + 4$, then the factor $n^2 + m^2$ in eq. (*), is:

$$n^2 + m^2 = 5(5a^2 + 5b^2 + 6a + 8b + 5)$$

which shows that $(n^2 + m^2)$, and therefore the product xyz is divisible by 5. Let the reader verify that for any choice of n and m, from eq. (***), **the product xyz is always divisible by 5.**

We have thus shown that the product xyz, in eq. (*), is divisible by 2 **and** by 3 **and** by 5, and therefore it is divisible by the product $2 \cdot 3 \cdot 5 = 30$.

Example 3-2-5: Find all the positive integers x, y and z which satisfy the system: $\{z^2 = x^2 + y^2, \quad z = x + 1\}$.

Solution: Substituting the expression of z, from the second equation, into the first equation, we find: $y^2 = 2x + 1$. It follows that y must be an **odd** number, (since if y were even, then its square would be even as well). Thus, the general form of y must be, $y = 2k + 1, k = 1,2,3, ...$ Then:

$$x = \frac{y^2 - 1}{2} = \frac{4k^2 + 4k + 1 - 1}{2} = 2k(k + 1)$$

$$z = x + 1 = 2k(k + 1) + 1$$

Summarizing, the solution of the system is:

$$\left\{ \begin{array}{l} x = 2k(k + 1) \\ y = 2k + 1 \\ z = 2k(k + 1) + 1 \end{array} \right\} \quad k = 1,2,3, ...$$

Example 3-2-6: Within the set of natural numbers solve the system:

$$\{2y^2 = x + 1 \qquad 2z^2 = x^2 + 1\}, \ x, y, z \in \mathbb{N}$$

Solution: From the first equation, $x = 2y^2 - 1$, it follows that x must be an odd number, so we may set, $x = 2k - 1, \ k = 1, 2, 3, ...$ Then, from the second equation, we find:

$$2z^2 = x^2 + 1 \Longrightarrow 2z^2 = (2k - 1)^2 + 1 = 4k^2 - 4k + 2 \Longrightarrow$$

$$z^2 = 2k^2 - 2k + 1 \Longrightarrow z^2 = (k - 1)^2 + k^2 \qquad (*)$$

a) If $k = 1$, then, $z = 1, x = 2 \cdot 1 - 1 = 1$, and $2y^2 = 1 + 1 = 2$, i.e. $y = 1$, (the negative solutions are not accepted, since x, y, z must be natural numbers). Thus, one solution of the system is: $(x, y, z) = (1,1,1)$.

b) If $k \neq 1$, (i.e. $k = 2, 3, 4, \ldots$), equation (*) is a Pythagorean equation. The solution of this equation is:

$$\{k - 1 = (m^2 - n^2)\lambda, \quad k = 2mn\lambda \quad z = (m^2 + n^2)\lambda\}, \qquad m, n\, \lambda \in \mathbb{N} \quad (**)$$

Since $k - 1 > 0$, it follows that $m > n$. From the first and the second equations in eq. (**), we have:

$$2mn\lambda - 1 = (m^2 - n^2)\lambda \Leftrightarrow (n^2 + 2mn - m^2)\lambda = 1 \qquad (***)$$

Since both factors in the left side of eq. (**) are integers, eq. (***) holds if

$$n^2 + 2mn - m^2 = 1 \; \textbf{ and } \; \lambda = 1 \qquad (****)$$

From the first equation in (****) we have:

$$n^2 + 2nm - m^2 - 1 = 0 \Leftrightarrow m(2n - m) = 1 - n^2 \qquad (*****)$$

We note that, if $m \neq 2$, then, eq. (*****) does not have an integer solution. To justify our assertion, we solve eq. (*****) for m,

$$m = \frac{-n^2 + 1}{2n - m}$$

and performing the algorithmic division, we find, (note that the numerator is a second degree polynomial in n and the denominator is a first degree polynomial in n):

$$m = -\left(\frac{n}{2} + \frac{m}{4}\right) + \frac{1 - \left(\frac{m}{2}\right)^2}{2n - m} \Leftrightarrow 4m = -(2n + m) + 4 \cdot \frac{1 - \left(\frac{m}{2}\right)^2}{2n - m} \quad (******)$$

Since $4m$ and $\{-(2n + m)\}$ are integers, the term $4 \cdot \{1 - (m/2)^2\}/(2n - m)$ must likewise be some integer, and this can occur only if the numerator is zero, i.e. if $\textbf{m = 2}$. For $m = 2$, from the first equation in (*****) we find: $n^2 + 4n - 5 = 0$, whose solutions are $n = -5$, (rejected, not a natural number), and $n = 1$, (accepted).

For $m = 2, n = 1$ and $\lambda = 1$, from the formulas in eq. (**), we find, $k = 4$, and then, $x = 2k - 1 = 8 - 1 = 7$, $z = (m^2 + n^2) = 2^2 + 1^2 = 5$, (from the third equation in eq. (**)), and finally, from $2y^2 = x + 1 = 7 + 1 = 8$, it follows $y = 2$, ($y = -2$ is rejected, not a natural number).

Summarizing, the solutions of the given system, within the set of natural numbers, are:

$$(x, y, z) = (1,1,1), (7,2,5)$$

Example 3-2-7: a) Show that if b is an integer, then $b^2 + 1 \neq mul. 3$, (i.e. $b^2 + 1$ cannot be a multiple of 3), **b)** If three integers x, y, z satisfy the equation $x^2 + y^2 = 3z^2$, then each one of these integers must be a multiple of 3, **c)** Show that the only integer solution of $x^2 + y^2 = 3z^2$ is the solution: $(x = 0, y = 0, z = 0)$, (the trivial solution).

Solution: a) For any integer b, we may write: $b = 3k + r$, where k is an integer and the remainder $r = 0, 1, 2$. Then:

$$b^2 + 1 = 9k^2 + 6kr + r^2 + 1 = 3(3k^2 + 2kr) + r^2 + 1 \qquad (*)$$

For $r = 0$, $b^2 + 1 = 3(3k^2) + 1 \neq mul. 3$.

For $r = 1$, $b^2 + 1 = 3(3k^2 + 2k) + 2 \neq mul. 3$.

For $r = 2$, $b^2 + 1 = 3(3k^2 + 4k) + 5 = 3(3k^2 + 4k + 2) - 1 \neq mul. 3$.

Thus, in all cases, $b^2 + 1 \neq \boldsymbol{mul. 3}$.

b) Let $x = 3k + r$, where $k \in \mathbb{Z}$ and $r \in \{0,1,2\}$. In this case:

$$x^2 + y^2 = 3z^2 \Longrightarrow (3k + r)^2 + y^2 = 3z^2 \Longrightarrow$$

$$y^2 + r^2 = 3 \underbrace{(z^2 - 3k^2 - 2kr)}_{\lambda} = 3\lambda, \quad \lambda \in \mathbb{Z} \qquad (**)$$

If $r = 1$, eq. (**) implies $y^2 + 1 = 3\lambda$, which **cannot** be true, as proved in part (a). If $r = 2$, eq.(**) implies $y^2 + 4 = 3\lambda$, or, $y^2 + 1 = 3(\lambda - 1)$, which again, by virtue of part (a) **cannot** be true. Thus the only possibility is $r = 0$, and then $x = 3k = mul. 3$. Similarly, we show that y must be a multiple of 3, i.e. $y = 3\lambda = mul. 3$, and then, from $x^2 + y^2 = 3z^2$, it follows that $z^2 = 3(k^2 + \lambda^2)$, i.e. $3 / z^2$, i.e. $3 / z$, i.e. $z = mul. 3$, and the proof has been completed.

c) If $(x, y, z) \neq (0,0,0)$ were a solution of $x^2 + y^2 = 3z^2$, then, as proved in part (b), x, y, z should all be multiples of 3. Let d be the Greatest Common

Divisor of x, y, z. If we call: $a = x/d, b = y/d, c = z/d$, then the G.C.D of a, b, c is 1, i.e. $(a, b, c) = 1$. Equation $x^2 + y^2 = 3z^2$ implies that $a^2 + b^2 = 3c^2$, which by virtue of part (b), implies that a, b, c should all be multiples of 3, contrary to our assumption that $(a, b, c) = 1$. Thus, we must, necessarily have, $(x, y, z) = (0,0,0)$.

PROBLEMS

3-2-1) Show that if $(x = k, y = \lambda, z = m)$ is a solution of the equation $x^2 + y^2 = z^2$, then:

$$(x = 3m + 3k + 2\lambda, \qquad y = 6m + 2k + 6\lambda, \qquad z = 7m + 3k + 6\lambda)$$

is also a solution of the same equation.

3-2-2) Solve the Diophantine equation: $x^2 + 5y^2 = z^2$.

3-2-3) Find the integer solutions of the equations: **a)** $x^2 - y^2 = 1$, **b)** $x^2 - 3y^2 = z^2$.

Ans: a) $x = \pm 1, y = 0$, **b)** $x = (3n^2 + m^2)k, y = 2mnk, z = (3n^2 - m^2)k$

Hint: The given equation is equivalent to: $z^2 + 3y^2 = x^2$.

3-2-4) Find the integer solutions of the equation:

$$x^2 + y^2 + 2x - 6y + 6 = z^2 + 4z$$

Hint: The equation may be written as $(x + 1)^2 + (y - 3)^2 = (z + 2)^2$. If we set, $a = x + 1$, $b = y - 3$, $c = z + 2$, the equation takes the form: $a^2 + b^2 = c^2$, etc.

3-2-5) Find the general form of the integers x, y such that $x^2 + 2xy + 2y^2$ to be a perfect square. Repeat for $x^2 + 5y^2$.

(Ans: a) $x = (m^2 - n^2 - 2mn)k$, $y = 2mnk$, **b)** $x = (5n^2 - m^2)k, y = 2mnk$, $m, n, k \in \mathbb{Z}$).

3-2-6) If $x, y, z, a, b, c \in \mathbb{Z}$, and $a^2 = b^2 + c^2$ and $z^2 = x^2 + y^2$, show that one of the two numbers, $az + by + cx$ and $az + cy + bx$, is a perfect square and the other is twice a perfect square, (in eq. (3-2-6) assume solutions with $k = 1$).

3-2-7) Find the integer solutions of the equations: **a)** $x^2 + y^2 = 13$, and

b) $5x^2 + 9y^2 = 81$.

(Ans: a) $(x, y) = (\pm 2, \pm 3)$, **b)** $(x, y) = (0, \pm 3), (\pm 3, \pm 2)$).

Hint: From $5x^2 + 9y^2 = 81$, $81 - 9y^2 = 5x^2 \geq 0$, i.e. $9y^2 \leq 81$, i.e. $y^2 \leq 9$, i.e. $y \in \{-3, -2, -1, 0, 1, 2, 3\}$. We check which one of these values of y yields an integer x, etc.

3-3) The homogeneous equation: $ax^2 + bxy + cy^2 = 0$

If the first degree terms and the constant term, in the general form of the Diophantine equation (3-1-1), are missing, the equation is called **homogeneous**. We assume that $abc \neq 0$.

An obvious solution of $ax^2 + bxy + cy^2 = 0$ is $(x, y, z) = (0,0,0)$, (the trivial solution). We are interested for the non-trivial solutions.

Let us consider the equation $ax^2 + bxy + cy^2 = 0$. If we divide both sides by y^2, we find:

$$a \left(\frac{x}{y}\right)^2 + b\frac{x}{y} + c = 0$$

or, if we set $t = x/y$,

$$at^2 + bt + c = 0, \qquad t = \frac{x}{y} \qquad\qquad (3-3-1)$$

Equation (3-3-1) is quadratic in t, and in the general case, gives two solutions. Provided that t is **a rational number**, (a fraction), we find the ratio (x/y), and from this we find the general form of the sought for integer solutions x and y. The following examples illustrate the method.

Example 3-3-1: Solve the Diophantine equation $4x^2 - 4xy - 3y^2 = 0$.

Solution: Dividing both sides by y^2 we find:

$$4t^2 - 4t - 3 = 0, \qquad t = x/y \qquad (*)$$

Solving this equation we find

$$t_{1,2} = \frac{-(-4) \pm \sqrt{(-4)^2 - 4 \cdot 4 \cdot (-3)}}{2 \cdot 4} = \frac{4 \pm \sqrt{64}}{8} = \frac{4 \pm 8}{8} \Longrightarrow$$

$$t_1 = \frac{12}{8} = \frac{3}{2}, \quad t_2 = -\frac{4}{8} = -\frac{1}{2} \qquad (**)$$

When $t = 3/2$:

$$t = \frac{x}{y} = \frac{3}{2} \Longrightarrow \frac{x}{3} = \frac{y}{2} = k, \Longrightarrow \begin{cases} x = 3k \\ y = 2k \end{cases}, \quad k \in \mathbb{Z} - \{0\} \qquad (***)$$

When $t = -1/2$:

$$t = \frac{x}{y} = \frac{-1}{2} \Longrightarrow \frac{x}{-1} = \frac{y}{2} = \lambda \Longrightarrow \begin{cases} x = -\lambda \\ y = 2\lambda \end{cases}, \quad \lambda \in \mathbb{Z} - \{0\} \qquad (****)$$

Equations (***) and (****) constitute the general solution of the equation. For example, for $k = 1$ we find one solution, (from eq. (***)), $(x, y) = (3,2)$, for $\lambda = -2$ we find another solution, (from eq. (****)), $(x, y) = (2, -4)$, etc.

Example 3-3-2: Solve the Diophantine equation: $3x^2 - 7xy - 2y^2 = 0$.

Solution: Working as in ex. 3-3-1, we find the quadratic equation:

$$3t^2 - 7t - 2 = 0, \quad t = \frac{x}{y} \qquad (*)$$

$$t_{1,2} = \frac{7 \pm \sqrt{(-7)^2 - 4 \cdot 3 \cdot (-2)}}{2 \cdot 3} = \frac{7 \pm \sqrt{73}}{6} \qquad (**)$$

Since x and y are integers, t must be **a rational number**, and since $\sqrt{73}$ is not rational, the given equation does not have integer solutions.

PROBLEMS

3-3-1) Solve the Diophantine equation: $x^2 + 2xy - 15y^2 = 0$.

(Ans: $(x, y) = (3k, k)$ or $(5\lambda, -\lambda)$, $k, \lambda \in \mathbb{Z} - \{0\}$).

3-3-2) Solve the Diophantine equation:
$$3(x - 2)^2 - 5(x - 2)(y - 4) + 2(y - 4)^2 = 0$$

Hint: Set $a = x - 2$, $b = y - 4$.

3-3-3) Solve the Diophantine equation: $9x^2 - 24xy + 16y^2 = 0$.

(Ans: $x = 4k, y = 3k$, $k \in \mathbb{Z} - \{0\}$).

3-4) Pell's equation: $x^2 - ky^2 = 1$

a) The Diophantine equation $Ax^2 + By^2 + Cx + Dy = F$, (the cross term xy is missing), can always be transformed to an equation of the form: $BX^2 + AY^2 = G$. To justify our assertion, we transform the original equation, as follows:

$$Ax^2 + By^2 + Cx + Dy = F \iff$$

$$A\left(x^2 + \frac{C}{A}x\right) + B\left(y^2 + \frac{D}{B}y\right) = F \iff$$

$$A\left(\underbrace{x^2 + 2 \cdot \frac{C}{2A}x + \left(\frac{C}{2A}\right)^2}_{Perfect\ Square} - \frac{C^2}{4A^2}\right) + B\left(\underbrace{y^2 + 2 \cdot \frac{D}{2B}y + \left(\frac{D}{2B}\right)^2}_{Perfect\ Square} - \frac{D^2}{4B^2}\right) = F$$

which is simplified to the following:

$$A\left(x + \frac{C}{2A}\right)^2 + B\left(y + \frac{D}{2B}\right)^2 = F + \frac{C^2}{4A} + \frac{D^2}{4B} \iff$$

$$B(2Ax + C)^2 + A(2By + D)^2 = 4ABF + C^2B + D^2A \qquad (3-4-1)$$

If we call: $X = 2Ax + C$, $Y = 2By + D$ and $G = 4ABF + C^2B + D^2A$, equation (3-4-1) becomes, $BX^2 + AY^2 = G$.

The conclusion is that any second degree Diophantine equation, **with the cross term xy missing,** may **always** be transformed into an equation of the form:

$$ax^2 + by^2 = c, \quad a, b, c \in \mathbb{Z} \qquad (3-4-2)$$

where the integer coefficients a, b, c are determined properly, in terms of the integer coefficients of the original equation.

b) Depending on the coefficients a, b and c, eq. (3-4-2) may have or may not have integer solutions. For example:

1) The equation $2x^2 + 3y^2 = -1$, does **not** have any solution.

2) The equation $5x^2 + 7y^2 = 0$ has **a unique** solution $(x, y) = (0,0)$.

3) To solve the equation $3x^2 + 4y^2 = 43$, we write this equation as:

$$3x^2 = 43 - 4y^2 \geq 0, \quad i.e. \quad y^2 \leq 43/4 \cong 10.7 \qquad (*)$$

The non negative **integer** values of y which satisfy the inequality in (*), are: $y = 0, 1, 2, 3$. If $y = 0$, the equation becomes: $3x^2 = 43$, which is **not** satisfied by an integer x. If $y = 1$, the equation becomes: $3x^2 = 43 - 4 = 39$, i.e. $x^2 = 13$, which, again, does **not** have integer solution.

If $y = 2$, the equation becomes: $3x^2 = 43 - 16 = 27$, i.e. $x^2 = 27/3 = 9$, i.e. $x = 3$. Thus $(x, y) = (3, 2)$ is one solution of the equation.

If $y = 3$, then $3x^2 = 43 - 36 = 7$, which does **not** have integer solutions.

Thus the non negative integer solution of the equation is $(x, y) = (3,2)$. Since in the original equation, x and y are raised to **an even power**, if (x, y) is one solution, then any combination of $(\pm x, \pm y)$ will also be a solution. Thus, the general solution of the equation $3x^2 + 4y^2 = 43$ is:

$$(x, y) = (3,2), (3, -2), (-3,2), (-3, -2)$$

4) In case where $a = k^2, b = -\lambda^2, \ k, \lambda \in \mathbb{Z} - \{0\}$, equation (3-4-2) becomes:

$$(kx)^2 - (\lambda y)^2 = c \iff (kx + \lambda y)(kx - \lambda y) = c \qquad (**)$$

This type of equation was studied in section 3-1 (2), (see Ex. 3-1-3).

c) Pell's equation: If $a = 1$, $b = -k$ ($k = 2,3,4, ...$) and $c = 1$, eq. (3-4-2) becomes:

$$x^2 - ky^2 = 1 \qquad \pmb{Pell's\ Equation} \qquad (3-4-3)$$

This equation is known as the "**Pell's equation**". We do not consider here the case $k = \lambda^2$, $\lambda = 1,2,3, ...$, since, in this case, eq. (3-4-3) becomes: $x^2 - (\lambda y)^2 = 1$, i.e. $(x - \lambda y)(x + \lambda y) = 1$, and this equation was studied in section 3-1 (2).

Thus the cases of interest, in Pell's equation, are the ones where **k is a positive integer and not a perfect square**. For example:

$$x^2 - 3y^2 = 1, \ \ or, \ \ x^2 - 7y^2 = 1, \ \ or, \ \ x^2 - 15y^2 = 1$$

If a Pell's equation has one integer solution (x_0, y_0), then, as it can be shown, it will have an infinite number of solutions. The method of solution of Pell's equation, (attributed to **Lagrange**), is illustrated in the following example.

Example 3-4-1: Solve Pell's equation: $x^2 - 3y^2 = 1$.

Solution: We note that $(x_0, y_0) = (2,1)$ is one solution of the equation. We write Pell's equation as a product of two factors, as follows:

$$\left(x + \sqrt{3}y\right)\left(x - \sqrt{3}y\right) = 1 \qquad (*)$$

If we square both sides of eq. (*) we find:

$$\left(x + \sqrt{3}y\right)^2 \left(x - \sqrt{3}y\right)^2 = 1 \Leftrightarrow$$

$$\{x^2 + 3y^2 + 2\sqrt{3}xy\}\{x^2 + 3y^2 - 2\sqrt{3}xy\} = 1 \Leftrightarrow$$

$$(x^2 + 3y^2)^2 - 3 \cdot 2^2 \cdot (xy)^2 = 1 \Leftrightarrow$$

$$(x^2 + 3y^2)^2 - 3(2xy)^2 = 1 \qquad (**)$$

If (x_0, y_0) is one, initial solution of Pell's equation, i.e. if $\pmb{x_0^2 - 3y_0^2 = 1}$, then, as shown in eq. (**), we must have:

$$(x_0{}^2 + 3y_0{}^2)^2 - 3(2x_0y_0)^2 = 1 \qquad\qquad (***)$$

This equation shows that $x_1 = x_0{}^2 + 3y_0{}^2$ and $y_1 = 2x_0y_0$ is **another solution** of Pell's equation, since $x_1{}^2 - 3y_1{}^2 = 1$, $(x_1, y_1$ **satisfies Pell's equation**). Thus, starting with an initial solution (x_0, y_0), we find another solution $x_1 = x_0{}^2 + 3y_0{}^2$ and $y_1 = 2x_0y_0$.

Working similarly, considering (x_1, y_1) as an initial solution, we find another solution, $(x_2, y_2) = \left(x_1{}^2 + 3y_1{}^2, 2x_1y_1\right)$, then, starting with (x_2, y_2) we find another solution $(x_3, y_3) = \left(x_2{}^2 + 3y_2{}^2, 2x_2y_2\right)$, etc.

In summary, one set of solutions of Pell's equation, are obtained from the following recursive formulas, (with an initial solution (x_0, y_0) to be known):

$$\{x_n = x_{n-1}{}^2 + 3y_{n-1}{}^2 \qquad y_n = 2x_{n-1}y_{n-1}\}, \qquad n = 1, 2, 3, \dots \quad (****)$$

In our example, with $x_0 = 2$ and $y_0 = 1$, we find:

$$\begin{Bmatrix} x_1 = 7 \\ y_1 = 4 \end{Bmatrix}, \quad \begin{Bmatrix} x_2 = 97 \\ y_2 = 56 \end{Bmatrix}, \quad \begin{Bmatrix} x_3 = 18817 \\ y_3 = 10864 \end{Bmatrix}, \quad etc \qquad (*****)$$

Comments: 1) As we see, **knowing one initial solution (x_0, y_0) of Pell's equation, we can find an infinite number of solutions**. With a different initial solution, for instance, $(x_0, y_0) = (-2, 1)$, in the previous example, we would find another, infinite set of solutions: $(x_1, y_1) = (7, -4)$, $(x_2, y_2) = (97, -56)$, etc.

2) In our example, we started with eq. (*) and squared both sides. Lagrange's method works equally well if we raise both sides to the third, or to the fourth, or to the fifth power, etc. In Example 3-4-2, we solve the same equation by raising both sides to the third power, and as we shall see, we obtain another, infinite set of solutions.

3) In general, writing Pell's equation $x^2 - ky^2 = 1$ as

$$\left(x - \sqrt{k}y\right)\left(x + \sqrt{k}y\right) = 1$$

and raising both sides to the nth power, $(n = 2, 3, 4, \dots)$, we find:

$$\left(x + \sqrt{k}\, y\right)^{n}\left(x - \sqrt{k}\, y\right)^{n} = 1$$

or, equivalently,

$$\left(X - \sqrt{k}\, Y\right)\left(X + \sqrt{k}\, Y\right) = 1 \Leftrightarrow X^2 - kY^2 = 1 \qquad (******)$$

where:

$$X = x^n + \frac{n(n-1)}{1 \cdot 2} x^{n-2} \cdot 3y^2 + \frac{n(n-1)(n-2)(n-3)(n-4)}{1 \cdot 2 \cdot 3 \cdot 4} x^{n-4} \cdot 3^2 y^4 + \cdots$$

$$Y = nx^{n-1}y + \frac{n(n-1)(n-2)}{1 \cdot 2 \cdot 3} x^{n-3} \cdot 3y^3 + \cdots$$

From eq. (******) it follows that, if (x, y) is one solution of Pell's equation, then, (X, Y) is also a solution, (since it satisfies Pell's equation), and since n can take on an infinite number of values, ($n = 2,3,4, ...$), we conclude that, if Pell's equation has one solution (x, y), then it will have an infinite number of solutions, (one for each n).

Example 3-4-2: Solve Pell's equation $x^2 - 3y^2 = 1$, by raising both sides of eq. (*) in example 3-4-1, to the third power.

Solution: In the previous example we found the solution of this equation for $n = 2$, and with initial solution $(x_0, y_0) = (2,1)$. Now we shall solve the same equation by raising both sides to the third power.

$$\left(x + \sqrt{3}y\right)\left(x - \sqrt{3}y\right) = 1 \Rightarrow \left(x + \sqrt{3}y\right)^{3}\left(x - \sqrt{3}y\right)^{3} = 1 \Rightarrow$$

$$\left(x^3 + 3x^2 \cdot \sqrt{3}y + 3x\left(\sqrt{3}y\right)^{2} + \left(\sqrt{3}y\right)^{3}\right)\left(x^3 - 3x^2 \cdot \sqrt{3}y + 3x\left(\sqrt{3}y\right)^{2}\right.$$
$$\left. - \left(\sqrt{3}y\right)^{3}\right) = 1 \Rightarrow$$

$$\{x^3 + 9xy^2 + \sqrt{3} \cdot (3x^2y + 3y^3)\}\{x^3 + 9xy^2 - \sqrt{3} \cdot (3x^2y + 3y^3)\} = 1 \Rightarrow$$

$$(x^3 + 9xy^2)^2 - 3(3x^2y + 3y^3)^2 = 1 \qquad (*)$$

Equation (*) shows that, if (x, y) is one solution of Pell's equation, then

$$\{X = x^3 + 9xy^2 \qquad Y = 3x^2y + 3y^3\} \qquad (**)$$

is also a solution of the same equation.

For example, from the initial solution $(x, y) = (2,1)$, we find another solution,

$$
\begin{cases}
X = 2^3 + 9 \cdot 2 \cdot 1^2 = 26 \\
Y = 3 \cdot 2^2 \cdot 1 + 3 \cdot 1^3 = 15
\end{cases}
$$

(Indeed, check by direct calculations that: $26^2 - 3 \cdot 15^2 = 1$).

From this solution we may find another one, by means of the formulas in eq. (**),

$$
\begin{cases}
X = 26^3 + 9 \cdot 26 \cdot 15^2 = 70226 \\
Y = 3 \cdot 26^2 \cdot 15 + 3 \cdot 15^3 = 40545
\end{cases}
$$

(Check that: $70226^2 - 3 \cdot 40545^2 = 1$).

Example 3-4-3: Solve Pell's equation: $x^2 - 5y^2 = 1$.

Solution: a) An initial solution is not given. In order to find an initial solution (x_0, y_0) we think as follows: Since the difference $x^2 - 5y^2 = 1$, (odd number), **one of the numbers x and y must be odd and the other must be even**, (since if both were odd **or** both were even, then their difference would be even). Let us suppose that x is odd. We try to see if there exists a solution, assuming that $x = 2y + 1$, (odd). In this case, the original equation becomes:

$$4y^2 + 4y + 1 - 5y^2 = 1 \Leftrightarrow 4y - y^2 = 0 \Leftrightarrow y(4 - y) = 0$$

It follows that $y = 4$, ($y = 0$ gives the trivial solution $(x, y) = (1,0)$). With $y = 4$, the associated $x = 9$. Thus, one solution of the equation, (an initial solution), is $(x_0, y_0) = (9, 4)$, (indeed, $9^2 - 5 \cdot 4^2 = 81 - 80 = 1$).

b) Working as in example 3-4-1, and with an initial solution $(x_0, y_0) = (9,4)$, we find that the solutions of the equation are given by the formulas:

$$\{x_n = x_{n-1}^2 + 5y_{n-1}^2 \qquad y_n = 2x_{n-1}y_{n-1}\}, \qquad n = 1, 2, 3, \ldots \qquad (*)$$

Thus, for $n = 1$, we find: $(x_1, y_1) = (161,72)$, for $n = 2$ we find: $(x_2, y_2) = (51841, 23184)$, etc.

Example 3-4-4: Solve the Diophantine equation:

$$x^2 - 5y^2 - 4x - 20y = 17$$

Solution: We have:

$$x^2 - 5y^2 - 4x - 20y = 17 \Leftrightarrow$$

$$(x^2 - 4x + 4) - 4 - 5(y^2 + 4y + 4) + 20 = 17 \Leftrightarrow$$

$$(x^2 - 4x + 4) - 5(y^2 + 4y + 4) = 17 + 4 - 20 \Leftrightarrow$$

$$(x - 2)^2 - 5(y + 2)^2 = 1 \qquad (*)$$

If we set: $a = x - 2, b = y + 2$, eq. (*) becomes: $a^2 - 5b^2 = 1$. This equation was solved in example 3-4-3, where we found:

$$\begin{cases} a_0 = 9 \\ b_0 = 4 \end{cases} \Leftrightarrow \begin{cases} x_0 - 2 = 9 \\ y_0 + 2 = 4 \end{cases} \Leftrightarrow \begin{cases} x_0 = 11 \\ y_0 = 2 \end{cases}$$

This is one solution. Another solution is found from $(a_1, b_1) = (161, 72)$, which gives:

$$\begin{cases} a_1 = 161 \\ b_1 = 72 \end{cases} \Leftrightarrow \begin{cases} x_1 - 2 = 161 \\ y_1 + 2 = 72 \end{cases} \Leftrightarrow \begin{cases} x_1 = 163 \\ y_1 = 70 \end{cases}$$

Similarly we may find (x_2, y_2), etc.

Example 3-4-5: Find two consecutive integers such the difference of their cubes is the square of another integer.

Solution: Let y and x be two integers such that: $(y + 1)^3 - y^3 = x^2$. Then,

$$y^3 + 3y^2 + 3y + 1 - y^3 = x^2 \Leftrightarrow x^2 = 3y^2 + 3y + 1 \qquad (*)$$

Since the cross term xy is missing, we may convert this equation to the following, (according to the method described in section 3-4 (a)):

$$(2x)^2 - 3(2y + 1)^2 = 1 \qquad (**)$$

The details in the derivation are left in Pr. 3-4-1.

If we set: $a = 2x$, $b = 2y + 1$, eq. (**) assumes the form:

$$a^2 - 3b^2 = 1 \qquad (***)$$

This is Pell's equation, and it was solved in examples 3-4-1 and 3-4-2. Note that in the solution set obtained in example 3-4-1, x is odd and y is even, while in the solution set of the same equation, in example 3-4-2, x is even and y is odd. Since in eq. (***), a is even ($a = 2x$) and b is odd ($b = 2y + 1$), we choose the solutions obtained in example 3-4-2, i.e.

$$(a, b) = (2,1), (26,15), (70226,40545), ...$$

and the associated x and y are found to be, ($x = a/2$, $y = (b - 1)/2$):

$$(x, y) = (1,0), (13,7), (35113,20272), ... \qquad (****)$$

We may check that the found values if x and y satisfy the requirements of the problem. For example:

For the pair $(x, y) = (1,0)$: $(0 + 1)^3 - 0^3 = 1^2$,

For the pair $(x, y) = (13,7)$: $(7 + 1)^3 - 7^3 = 13^2$,

For the pair $(x, y) = (35113,20272)$: $(20272 + 1)^3 - 20272^3 = 35113^2$

Of course, there are infinitely many other pairs that satisfy the requirements of the problem.

PROBLEMS

3-4-1) In example 3-4-5, starting with eq. (*) derive eq. (**).

3-4-2) Solve the Diophantine equation: $x^2 - 3y^2 + 8x + 36y = 93$.

(Ans: $(x, y) = (3,10), (93,62), ...$).

Hint: Show that the given equation may be transformed to the following: $(x + 4)^2 - 3(y - 6)^2 = 1$, then, set $a = x + 4$, $b = y - 6$, etc, (see Ex. 3-4-4).

3-4-3) Find one set of solutions Pell's equation: $x^2 - 2y^2 = 1$, by writing this equation as $(x + y\sqrt{2})(x - y\sqrt{2}) = 1$, and squaring both sides, (note that $(x_0, y_0) = (3,2)$ is an initial solution of the equation).

(**Ans:** $(x, y) = (17,12), (577,408), (665857,470832), ...).$

3-4-4) Find another set of solution of the same equation in problem 3-4-3, starting with $(x + y\sqrt{2})(x - y\sqrt{2}) = 1$, and raising both sides to the third power.

(**Ans:** $(x, y) = (99,70), (3880899,2744210), ...).$

Hint: See example 3-4-2.

CHAPTER 4: HIGHER DEGREE DIOPHANTINE EQUATIONS (EXAMPLES)

There is no general theory for solving Diophantine equations of degree higher than two. This problem is highly individualistic, and certainly, much harder than the problem of solving a second degree Diophantine equation. In this chapter we shall consider a few examples which highlight some general guidelines and provide a general method of approach.

Example 4-1: Find the integer and positive numbers x and y that satisfy the equation: $x^4 + 4y^4 = p$, where $p > 0$ is a prime number.

Solution: The given equation may be written as:

$$\underbrace{x^4 + 4y^4 + 4x^2y^2}_{(x^2+(2y)^2)^2} - 4x^2y^2 = p \Leftrightarrow (x^2 + (2y)^2)^2 - (2xy)^2 = p \Leftrightarrow$$

$$\{x^2 + (2y)^2 + 2xy\}\{x^2 + (2y)^2 - 2xy\} = p \Leftrightarrow$$

$$\left\{\underbrace{x^2 + 2xy + y^2}_{(x+y)^2} + y^2\right\}\left\{\underbrace{x^2 - 2xy + y^2}_{(x-y)^2} + y^2\right\} = p \Leftrightarrow$$

$$\{(x + y)^2 + y^2\}\{(x - y)^2 + y^2\} = p \qquad (*)$$

Since p is **a prime number**, its only divisors, are p and 1. Therefore, the smallest number in eq. (*), which is $(x - y)^2 + y^2$, must be equal to 1, and the other number must be equal to p, i.e.

$$\begin{cases}(x + y)^2 + y^2 = p \\ and \\ (x - y)^2 + y^2 = 1\end{cases} \qquad (**)$$

Since x and y are integers, the second equation in (**) implies that, either $(x - y) = 1$ and $y = 0$, **or**, $(x - y) = 0$ and $y = 1$.

In the first case, from the second equation in (**), with $y = 0$ it follows that $x = 1$, and then, from the first equation in (**), we find $1^2 = 1 = p$, which is not true, since $p \geq 2$.

In the second case, it follows, $x = y = 1$, and therefore, from the first equation in (**), we have: $(1 + 1)^2 + 1^2 = p$, i.e. $p = 5$.

Summarizing, the two positive integers x and y, which satisfy the equation are $x = 1, y = 1$, and the equation has a solution only when $p = 5$. If p is a prime $\neq 5$, the equation does not have integer solution.

Example 4-2: Find the integer solutions of the system:

$$2y^3 = x^3 + z^3, \qquad 2y = x + z$$

Solution: From the second equation, $y = (x + z)/2$, and substituting into the first equation we find:

$$2\left(\frac{x + z}{2}\right)^3 = x^3 + z^3 \Leftrightarrow (x + z)^3 = 4(x^3 + z^3) \Leftrightarrow$$

$$x^3 + 3x^2z + 3xz^2 + z^3 = 4x^3 + 4z^3 \Leftrightarrow$$

$$3(x^3 + x^2z + xz^2 + z^3) = 0 \Leftrightarrow x^3 + x^2z + xz^2 + z^3 = 0 \Leftrightarrow$$

$$x^2(x + z) + z^2(x + z) = 0 \Leftrightarrow (x^2 + z^2)(x + z) = 0 \qquad (*)$$

Equation (*) is satisfied if $x^2 + z^2 = 0$, i.e. if $x = z = 0$, and then $y = 0$, or, if $(x + z) = 0$, and then $y = 0$, i.e. if $x = -z$ and $y = 0$. Thus the solutions of the system are: $(x, y, z) = (0,0,0)$, or, $(x, y, z) = (-k, 0, k), \ k \in \mathbb{Z}$.

Example 4-3: Find the positive integers x, y, z which are relatively prime and satisfy the equation: $x^2 + y^2 = z^4$.

Solution: The given equation is written as: $x^2 + y^2 = (z^2)^2$, and its solution, according to eq. (3-2-6) are given by the formulas:

$$\{x = m^2 - n^2, \quad y = 2mn, \quad z^2 = m^2 + n^2\} \qquad (*)$$

Note that the integer k which appears in eq. (3-2-6) is taken here to be 1, since if $k \neq 1$, the integers x, y, z would have a common factor $k \neq 1$, and therefore they would not be relatively prime.

The third equation in (*) is itself another Pythagorean equation. Its solutions are given by the formulas:

$$\{m = k^2 - \lambda^2, \quad n = 2k\lambda, \quad z = k^2 + \lambda^2\} \tag{**}$$

Substituting these expressions of m and n, (in terms of k and λ), into equation (*), we find:

$$\{x = \pm(k^4 - 6k^2\lambda^2 + \lambda^4), \quad y = \pm 4k\lambda(k^2 - \lambda^2), \quad z = k^2 + \lambda^2\}$$

with one of the two numbers k and λ being odd and the other being even, (since x, y, z are relatively prime).

Example 4-4: Find the integer solutions of the equation: $x^3 - 4x = 3 + 4y$.

Solution: Solving the equation for y, we find:

$$y = \frac{x^3 - 3 - 4x}{4} = \frac{x^3 - 3}{4} - x \tag{*}$$

For x integer, the corresponding y will be integer, if and only if the term $(x^3 - 3)/4$ is some other integer, i.e. **if 4 divides exactly** $(x^3 - 3)$.

Any integer x can be written as: $x = 4k + r$, where $r \in \{0,1,2,3\}$ and $k \in \mathbb{Z}$

a) For $r = 0, x = 4k$:

$$\frac{x^3 - 3}{4} = \frac{(4k)^3 - 3}{4} = 4^2 k^3 - \frac{3}{4} \quad (Not\ integer)$$

b) For $r = 1, x = 4k + 1$:

$$\frac{x^3 - 3}{4} = \frac{(4k + 1)^3 - 3}{4} = \frac{(4k)^3 + 3 \cdot (4k)^2 + 3 \cdot (4k) + 1 - 3}{4} \Longrightarrow$$

$$\frac{x^3 - 3}{4} = 4^2 k^3 + 3 \cdot 4k^2 + 3k - \frac{2}{3} \quad (Not\ integer)$$

c) For $r = 2, x = 4k + 2$:

$$\frac{x^3 - 3}{4} = \frac{(4k + 2)^3 - 3}{4} = \frac{(4k)^3 + 3 \cdot (4k)^2 \cdot 2 + 3 \cdot (4k) \cdot 2^2 + 2^3 - 3}{4} \Longrightarrow$$

$$\frac{x^3 - 3}{4} = 4^2 k^3 + 3 \cdot 4k^2 \cdot 2 + 3k \cdot 2^2 + \frac{5}{3} \quad (Not\ integer)$$

d) For $r = 3$, $x = 4k + 3$:

$$\frac{x^3 - 3}{4} = \frac{(4k+3)^3 - 3}{4} = \frac{(4k)^3 + 3 \cdot (4k)^2 \cdot 3 + 3 \cdot (4k) \cdot 3^2 + 3^3 - 3}{4} \Rightarrow$$

$$\frac{x^3 - 3}{4} = 4^2 k^3 + 3 \cdot 4k^2 \cdot 3 + 3k \cdot 3^2 + 6$$
$$= 16k^3 + 36k^2 + 27k + 6 \qquad (Integer)$$

Thus, $(x^3 - 3)/4$ is an integer only when $x = 4k + 3$, $k \in \mathbb{Z}$. The corresponding y, as obtained from eq. (*) is $y = 16k^3 + 36k^2 + 27k + 3$.

Thus, the integer solutions of the equation are given by the formulas:

$$\left\{ \begin{array}{l} x = 4k + 3 \\ y = 16k^3 + 36k^2 + 23k + 3 \end{array} \right\} \quad k \in \mathbb{Z}$$

For example, for $k = 0$, one solution is $(x, y) = (3,3)$, for $k = 1$ we find another solution $(x, y) = (7,78)$, etc.

Example 4-5: Find the integer and positive solutions of the equation:

$$200x^7 = 3y^3$$

Solution: In terms of its prime divisors, $200 = 2^3 \cdot 5^2$, and therefore,

$$2^3 \cdot 5^2 x^7 = 3y^3 \qquad (*)$$

Since 2,3,5 are prime numbers, we may try a solution of the form

$$\{x = 2^a \cdot 3^b \cdot 5^c \quad and \quad y = 2^k \cdot 3^\lambda \cdot 5^\mu\}, \qquad a, b, c, k, \lambda, \mu \in \mathbb{Z} \quad (**)$$

where the **positive integers** a, b, c, k, λ, μ are to be determined so that equation (*) is satisfied. Substituting these expressions of x and y into eq. (*) we find:

$$2^{7a+3} \cdot 3^{7b} \cdot 5^{7c+2} = 2^{3k} \cdot 3^{3\lambda+1} \cdot 5^{3\mu}$$

This equality holds true, provided that:

$$\{7a + 3 = 3k, \quad 7b = 3\lambda + 1, \quad 7c + 2 = 3\mu\}$$

or, equivalently,

$$\{3k - 7a = 3, \quad 7b - 3\lambda = 1, \quad 3\mu - 7c = 2\} \qquad (***)$$

Each one of the equations in (***) is a Diophantine equation, which is solved as described in section 1-2. We find:

$$\begin{cases} k = 1 + 7t, \ a = 3t \\ \lambda = 2 + 7u, \ b = 1 + 3u \\ \mu = 3 + 7w, \ c = 1 + 3w \end{cases} \quad where \quad t, u, w \in \{0,1,2,3,\dots\} \qquad (****)$$

Note that all t, u, w must be zero or positive integers, since a, b, c, k, λ, μ must be positive integers. Then, from eq. (**) we have:

$$x = 2^a \cdot 3^b \cdot 5^c \implies$$

$$x = 2^{3t} \cdot 3^{1+3u} \cdot 5^{1+3w} = 15 \cdot 2^{3t} \cdot 3^{3u} \cdot 5^{3w} = 15 \cdot (2^t \cdot 3^u \cdot 5^w)^3 \quad (*****)$$

$$y = 2^k \cdot 3^\lambda \cdot 5^\mu \implies$$

$$y = 2^{1+7t} \cdot 3^{2+7u} \cdot 5^{3+7w} = 2 \cdot 3^2 \cdot 5^3 \cdot 2^{7t} \cdot 3^{7u} \cdot 5^{7w} \implies$$

$$y = 2250 \cdot (2^t \cdot 3^u \cdot 5^w)^7 \qquad (******)$$

With positive integers t, u, w, the product $2^t \cdot 3^u \cdot 5^w$ will be some integer, say m, i.e. $m = 2^t \cdot 3^u \cdot 5^w$, and thus, the solution of the given equation is:

$$\{x = 15m^3 \qquad y = 2250m^7\}, \quad m \in \mathbb{N}$$

Example 4-6: Find the positive integers x, y, z, with $x < y < z$, which satisfy the equation $3^x + 3^y + 3^z = 6831$.

Solution: In terms of its prime divisors, $6831 = 3^3 \cdot 11 \cdot 23$, and thus,

$$3^x + 3^y + 3^z = 3^3 \cdot 11 \cdot 23 \iff$$

$$3^x(1 + 3^{y-x} + 3^{z-y}) = 3^3 \cdot 11 \cdot 23 \qquad (*)$$

Since $x < y < z$, (by assumption), $(y - x) \in \mathbb{N}, (z - y) \in \mathbb{N}$, and therefore $(1 + 3^{y-x} + 3^{z-y})$ is some positive integer. Equation (*) implies, $x = 3$ and $(1 + 3^{y-x} + 3^{z-y}) = 11 \cdot 23 = 253$, i.e. $3^{y-3} + 3^{z-3} = 252 = 2^2 \cdot 3^2 \cdot 7$, i.e.

$$3^{y-3}(1 + 3^{z-y}) = 2^2 \cdot 3^2 \cdot 7 \qquad (**)$$

and since $z > y$, $1 + 3^{z-y}$ is some integer, and therefore eq. (**) implies, $y - 3 = 2$, i.e. $y = 5$, and $(1 + 3^{z-y}) = (1 + 3^{z-5}) = 2^2 \cdot 7 = 28$, i.e. $3^{z-5} = 27 = 3^3$, i.e. $z - 5 = 3$, i.e. $z = 8$.

The sought for solution of the equation is: $(x = 3, y = 5, z = 8)$.

Example 4-7: Find the integer numbers (x, y, z) which are relatively prime in pairs and satisfy the equation: $x^2 y + y^2 z + z^2 x = k\,xyz$, with $k \in \mathbb{N}$, (assume $xyz \neq 0$), and then determine k.

Solution: Let $(x = a, y = b, z = c)$ be a solution of the system, with $a, b, c \in \mathbb{Z} - \{0\}$. Then,

$$a^2 b + b^2 c + c^2 a = kabc \Leftrightarrow a^2 b + b^2 c = kabc - c^2 a \Leftrightarrow$$

$$b(a^2 + bc) = ac(kb - c) \qquad (*)$$

This equation shows that the number b must **necessarily** be either 1 or -1, i.e. $b = \pm 1$. Because, if we assume that $b \neq \pm 1$, we are led to a contradiction. Indeed, if we assume that $b \neq \pm 1$, then from eq. (*) it follows that b must divide the number $ac(kb - c)$, (the quotient of the division is $(a^2 + bc)$, integer), and since b is relatively prime to a and c, (by hypothesis), b must divide the number $(kb - c)$. However, since b divides kb, it will also divide the difference $\{kb - (kb - c)\} = c$, i.e. $b\,/\,c$, which cannot be true, since b and c are relatively prime. Thus, necessarily, $b = \pm 1$. Similarly, we show that $a = \pm 1$ and $c = \pm 1$. The solutions (a, b, c) of the equation are summarized in the following table, (in total, there are eight solutions):

Table 1: Solutions (a, b, c) of the equation:

a	1	1	1	1	-1	-1	-1	-1
b	1	1	-1	-1	1	1	-1	-1
c	1	-1	1	-1	1	-1	1	-1

For the triads $(a, b, c) = (1,1,1), (-1, -1, -1)$ it follows that $k = 3$, while for the other six triads, $k = -1$, (check it).

Example 4-8: Find the integer solutions of the system:

$$\{x^4 + z^4 = 97, \quad y^2 + z^2 = 10\}$$

Solution: a) Let us find first the integer, positive solutions of the system, $(x, y, z \in \mathbb{N})$. Solving the second equation for z^2 and substituting into the first, we find:

$$x^4 + (10 - y^2)^2 = 97 \Leftrightarrow x^4 + 100 - 20y^2 + y^4 = 97 \Leftrightarrow$$

$$y^4 - 20y^2 + x^4 + 3 = 0 \qquad (*)$$

If we set: $x^2 = \phi$, $y^2 = w$, eq. (*) becomes:

$$w^2 - 20w + \phi^2 + 3 = 0 \;\; (w > 0, \phi > 0) \qquad (**)$$

This is a quadratic equation in w. The discriminant of this equation must be a non negative number, (for w to be real), i.e.

$$(-20)^2 - 4 \cdot (\phi^2 + 3) \geq 0 \Longrightarrow 400 - 4\phi^2 - 12 \geq 0 \Longrightarrow$$

$$388 \geq 4\phi^2 > 0 \Longrightarrow 97 \geq \phi^2 > 0 \Longrightarrow \sqrt{97} \cong 9.84 \geq \phi > 0 \overset{(\phi = x^2)}{\Longrightarrow}$$

$$9.84 \geq x^2 > 0 \Longrightarrow \sqrt{9.84} \cong 3.13 \geq x > 0 \Longrightarrow x \in \{1,2,3\} \qquad (***)$$

If $x = 1$, the first equation of the system becomes: $1 + z^4 = 97$, i.e. $z = \sqrt[4]{96}$, not integer, and therefore $x = 1$ is rejected.

If $x = 2$, again from the first equation of the system, $2^4 + z^4 = 97$, or, $z^4 = 97 - 16 = 81$, i.e. $z = 3$. With $x = 1$ and $z = 3$, the second equation of the system yields $y = 1$. Thus one solution of the system is: $(x, y, z) = (2,1,3)$.

If $x = 3$, the first equation of the system becomes: $3^4 + z^4 = 97$, i.e. $z^4 = 97 - 81 = 16$, i.e. $z = \sqrt[4]{16} = 2$, and with this value of z the second equation of the system becomes, $y^2 + 4 = 10$, i.e. $y^2 = 6$, which does not even have integer solution. Thus, in **the set of natural numbers**, the system admits one solution, $(x, y, z) = (2,1,3)$.

b) Because all the unknowns are raised **to even powers**, the general solution of the system **in the set of integers**, is: $(x, y, z) = (\pm 2, \pm 1, \pm 3)$. Thus, $(-2, 1, -3)$ is one solution, $(2, 1, -3)$ is another solution, etc.

PROBLEMS

4-1) Within the set of integers $\mathbb{Z}$ solve the equation: $2x^4 + 3y^4 = 2037$.

(Ans: $(x, y) = (\pm 3, \pm 5)$).

Hint: $2037 - 3y^4 = 2x^4 \geq 0$, i.e. $2037 \geq 3y^4$, i.e. $2037/3 \geq y^4$, i.e. $y \in \{1, 2, 3, 4, 5\}$, etc. See Ex. 4-8.

4-2) Find the integer and positive solution of the equations:

$$a) \; 27x^3 - y^3 = 19 \qquad b) \; 8x^3 + y^3 = 9$$

(Use the identity: $a^3 \pm b^3 = (a \pm b)(a^2 \mp ab + b^2)$).

Hint: a) $(3x)^3 - y^3 = 19$, i.e. $(3x - y)(9x^2 + 3xy + y^2) = 19$, and since 19 is a prime number, this means: $\{3x - y = 1 \; \textbf{and} \; 9x^2 + 3xy + y^2 = 19\}$, or, $\{3x - y = 19 \; \textbf{and} \; 9x^2 + 3xy + y^2 = 1\}$, or, $\{3x - y = -1 \; \textbf{and} \; 9x^2 + 3xy + y^2 = -19\}$, or, $\{3x - y = -19 \; \textbf{and} \; 9x^2 + 3xy + y^2 = -1\}$.

b) $(2x)^3 + y^3 = 9$, i.e. $(2x + y)(4x^2 - 2xy + y^2) = 9$, which means: $\{2x + y = 9 \; \textbf{and} \; 4x^2 - 2xy + y^2 = 1\}$, or $\{2x + y = 3 \; \textbf{and} \; 4x^2 - 2xy + y^2 = 3\}$, or $\{2x + y = 1 \; \textbf{and} \; 4x^2 - 2xy + y^2 = 9\}$, or, $\{2x + y = -9 \; \textbf{and} \; 4x^2 - 2xy + y^2 = -1\}$, or, $\{2x + y = -3 \; \textbf{and} \; 4x^2 - 2xy + y^2 = -3\}$, or, $\{2x + y = -1 \; \textbf{and} \; 4x^2 - 2xy + y^2 = -9\}$.

4-3) Solve the Diophantine system:

$$16x^3 = 343y^4$$

(Ans: $x = 7k^4$, $y = 2k^3$, $k \in \mathbb{Z}$).

Hint: $16 = 2^4$, $343 = 7^3$, assume $x = 2^a \cdot 7^b$, $y = 2^k \cdot 7^\lambda$, See Ex. 4-5).

4-4) Assuming that $(x = a, y = b, z = c)$ is one solution of the equation $x^3 + y^3 = 9z^3$, show that:

$$\{x = a(a^3 + 2b^3), \quad y = -b(b^3 + 2a^3), \quad z = c(a^3 - b^3)\}$$

is also a solution.

4-5) Find the integers x and y, which are relatively prime and satisfy the equation:

$$\frac{y^2 - x^2}{y^3 - x^3} = \frac{12}{109}, \quad (x, y) = 1$$

(**Ans:** $(x, y) = (7,5), (5,7)$).

4-6) If $k, p \in \mathbb{N}$ and p is prime, solve the equation: $k^4 + 4 = p$.

(**Ans:** $k = 1, \ p = 5$).

4-7) Solve the Diophantine equation, $(x, y, z$ positive integers):

$$xy + yz + zx = xyz$$

(**Ans:** $(x, y, z) = \begin{cases} (3,3,3) \quad (4,4,2) \quad (4,2,4) \quad (2,4,4) \quad (6,3,2) \\ (3,2,6) \quad (2,6,3) \quad (3,6,2) \quad (6,2,3) \quad (2,3,6) \end{cases}$)

Hint: The given equation is equivalent to:

$$\frac{1}{x} + \frac{1}{y} + \frac{1}{z} = 1$$

MISCELLANEOUS PROBLEMS

1) Find the general form of fractions having the property: The sum of their squares is equal to 2.

(**Ans:** The fractions are of the form: x/z and y/z, where:

$$\begin{cases} 2x = (m^2 - n^2 + 2mn)k, \\ 2y = (m^2 - n^2 - 2mn)k, \\ \quad\ 2z = (m^2 + n^2)k \end{cases} \qquad m, n, k \in \mathbb{Z}$$

The integers m and n, **are either both even, or, both odd**. For example, for $k = 1, m = 4, n = 2$, the fractions are: $(14/10)$ and $(- 2/10)$. Notice that $(14/10)^2 + (- 2/10)^2 = 200/100 = 2$. For $k = 1, m = 3, n = 1$, we find: $(7/5)$ and $(1/5)$, $(7/5)^2 + (1/5)^2 = 50/25 = 2)$.

Hint: Let x/w and y/z be the two sought for fractions, in their lowest terms. This means that x and w are relatively prime, and similarly y and z are relatively prime, i.e. $(x, w) = 1, (y, z) = 1$. Then:

$$\left(\frac{x}{w}\right)^2 + \left(\frac{y}{z}\right)^2 = 2 \Longrightarrow x^2z^2 + y^2w^2 = 2z^2w^2 \Longrightarrow$$

$$y^2w^2 = 2z^2w^2 - x^2z^2$$

It follows that w^2 must divide $(2z^2w^2 - x^2z^2)$, and since it divides $2z^2w^2$, w^2 must divide x^2z^2 as well. However, since $(x, w) = 1$, w^2 must divide z^2, i.e. w must divide z. Similarly, we can show that z must divide w. The conclusion is that $z = \pm w$. Then,

$$\left(\frac{x}{w}\right)^2 + \left(\frac{y}{z}\right)^2 = 2 \xrightarrow{(z=\pm w)} x^2 + y^2 = 2z^2 \Longrightarrow 2x^2 + 2y^2 = 4z^2 \Longrightarrow$$

$$(x + y)^2 + (x - y)^2 = (2z)^2 \qquad\qquad (*)$$

If we set: $X = x + y$, $Y = x - y$, $Z = 2z$, eq. (*) becomes, $X^2 + Y^2 = Z^2$, which is a Pythagorean equation. Solving this equation we find X, Y, Z and then find x, y, z.

2) Solve the Diophantine system:

$$\{x + 2y + 3z = 6, \qquad 3x - 5y + z = 7\}$$

(**Ans:** $x = 4 - 17k, \ y = 1 - 8k, \ z = 11k, \ k \in \mathbb{Z}$).

3) Find all the integer and non negative numbers a and b which satisfy the equation: $2a + 3b = 20$.

(**Ans:** $(a, b) = (1,6), (4,4), (7,2), (10,0)$).

Hint: The general solution of the equation is: $a = 4 + 3k, b = 4 - 2k$. The nonnegative solutions, ($a \geq 0$ **and** $b \geq 0$), are obtained for $k = -1, 0, 1, 2$.

4) Find the integer solutions of the equation:

$$x^2 + 2xy + y^2 + y - x = 4$$

Ans: $(x, y) = (2\lambda^2 + \lambda - 2, -2\lambda^2 + \lambda + 2), (2\lambda^2 + 3\lambda - 1, -2\lambda^2 - \lambda + 2)$ where $\lambda \in \mathbb{Z}$.

Hint: $(x + y)^2 + (y - x) = 4$, or, if we set: $w = x + y, z = y - x$, $z = 4 - w^2$, whose integer solutions are: $w = k, z = 4 - k^2$ with $k \in \mathbb{Z}$. In terms of x, y: $2x = k - k^2 + 4, \ 2y = k + k^2 - 4$. It follows that $k - k^2 + 4$ and $k + k^2 - 4$ must **both** be multiples of 2, (since x and y are integers). For $k = 2\lambda$ (even), we find the first solution, and for $k = 2\lambda + 1$, (odd), we find the second solution.

5) Find the maximum value of c, for which the equation $7x + 9y = c$ has six integer and positive solutions.

(**Ans:** Maximum value of $c = 440$).

Hint: Let $(x_0 > 0, y_0 > 0)$ be **the first positive solution**. All other solutions of the equation can be expressed as: $\{x = x_0 + 9k, \ y = y_0 - 7k\}, \ k \in \mathbb{Z}$.

The six positive solutions of the equation, (since (x_0, y_0) is the first one), are:

$$\begin{cases} x_0 & x_0 + 9 & x_0 + 2 \cdot 9 & x_0 + 3 \cdot 9 & x_0 + 4 \cdot 9 & x_0 + 5 \cdot 9 \\ y_0 & y_0 - 7 & y_0 - 2 \cdot 7 & y_0 - 3 \cdot 7 & y_0 - 4 \cdot 7 & y_0 - 5 \cdot 7 \end{cases}$$

Since $y_0 - 5 \cdot 7 > 0$, $y_0 > 35$. The next value of y, which is $y_0 - 6 \cdot 7$ must be negative, i.e. $y_0 - 6 \cdot 7 < 0$, i.e. $y_2 < 42$. Also, the term $x_0 - 9$ must be negative, i.e. $0 < x_0 < 9$. Summarizing, we have found that:

$$\begin{cases} 0 < x_0 < 9 \\ 35 < y_0 < 42 \end{cases} \Longrightarrow \begin{cases} 0 < 7x_0 < 7 \cdot 9 \\ 35 \cdot 9 < 9y_0 < 42 \cdot 9 \end{cases}$$

If we add the two inequalities, we find:

$$35 \cdot 9 < \underbrace{7x_0 + 9y_0}_{c} < 7 \cdot 9 + 42 \cdot 9 \Longrightarrow 315 < c < 441$$

This inequality shows that the maximum integer value of c is 440.

6) Solve for Diophantine equation: $2x^2 + 7xy + 3y^2 - 5y = 2$.

(Ans: $(x, y) = (k, 2 - 2k)$, $(3k - 1, -k)$, $k \in \mathbb{Z}$**).**

Hint: If we consider the given equations as a quadratic in x, we find that its descriminant is a perfect square, and therefore we may factor the given equation as follows:

$$2x^2 + 7xy + 3y^2 - 5y - 2 = (2x + y - 2)(x + 3y + 1) = 0$$

Thus, the original equation is split into two linear equations, etc.

7) Find the integer solutions of the system:

$$\{x(y + z) = 2yz, \quad x + y + z = 11, \quad x^2 + y^2 + z^2 = 49\}$$

(Ans: $(x, y, z) = (3,2,6), (3,6,2)$**).**

Hint: If we square the second equation and then use the first and the third, we find: $3yz = 36$, i.e. $yz = 12$, and since y, z are integers, the solutions of this equation are: $y = k, z = 12/k$, where: $k = \pm 1, \pm 2, \pm 3, \pm 4, \pm 6, \pm 12$, (the divisors of 12). For each pair (y, z) we find the associated $x = 11 - y - z$, (from the second equation). From the 12 triads (x, y, z) we find, we eliminate those that do not satisfy the first and the third equation.

8) Solve the Diophantine equations:

$$a)\ 9x^2 - 25y^2 = 44 \qquad b)\ 9x^2 + 4y^2 = z^2$$

Hint: a) $(3x)^2 - (5y)^2 = 44$, i.e. $(3x - 5y)(3x + 5y) = 44$, and this implies that $(3x - 5y) = d$ **and** $(3x + 5y) = 44/d$, where d is a divisor of 44, $(d = \pm1, \pm2, \pm4, \pm11, \pm22, \pm44)$, **b)** $(3x)^2 + (2y)^2 = z^2$, or if we set: $X = 3x, Y = 2y, X^2 + Y^2 = z^2$, the solution of which is:

$$\{X = (m^2 - n^2)k, \quad Y = 2mnk \quad z = (m^2 + n^2)k\}$$

Since $X = 3x$, i.e. X must be divisible by 3, $k = 3\lambda$, or else, 3 must divide $(m^2 - n^2) = (m - n)(m + n)$, i.e. 3 must divide either $(m - n)$ or $(m + n)$. In the first case, $m = 3\lambda + n$, in the second case, $m = 3\lambda - n$, etc. Thus, for $m = 3\lambda + n$, we find one family of solutions, (x, y, z):

$$\{x = \lambda(3\lambda + 2n)k, \quad y = (3\lambda + n)nk, \quad z = (9\lambda^2 + 2n^2 + 6\lambda n)k\}$$

For example, for $\lambda = 2, n = 3, k = 1$, we find $(x = 24, y = 27, z = 90)$, which is indeed a solution of the original equation, etc.

9) Find the integer and positive solutions of the system:

$$\{x + y + z = 43 \qquad 10x + 5y + 2z = 229\}$$

(Ans: $(x, y, z) = (1 + 3k, 45 - 8k, -3 + 5k), \quad k \in \mathbb{Z})$. This is the general solution. For the positive solutions, we must determine k so that $x > 0$ **and** $y > 0$ **and** $z > 0$. The three inequalities are satisfied simultaneously, only for $k = 1, 2, 3, 4, 5$.

10) If $2x + 3y = mul.\ 17$, show that $9x + 5y = mul.\ 17$, and conversely.

Hint: One solution of the equation $2x + 3y = 17k$, is $x_0 = k, y_0 = 5k$, and therefore its general solution is: $x = k + 3\lambda, \ y = 5k - 2\lambda, \ k, \lambda \in \mathbb{Z}$.

11) Find the integer solutions (x, y) which are common to the two equations: $2x + 5y = 11$ and $2x^2 + 3xy - y^2 + 7x - y = 46$.

(Ans: $(x, y) = (3,1))$.

12) If p is a prime number, such that $10 < p < 40$, solve the Diophantine equation: $3x^2 + xy - 2y^2 = p$.

(**Ans:** $(x, y) = (7,10), (15,22), (3, -2), (5, -4)$).

Hint: $3x^2 + xy - 2y^2 = (x + y)(3x - 2y) = p$, and since p is a prime number, $\{x + y = p,\ 3x - 2y = 1\}$, or, $\{x + y = 1,\ 3x - 2y = p\}$, i.e. $\{x = (2p + 1)/5, y = (3p - 1)/5\}$, or, $\{x = (p + 2)/5, y = (3 - p)/5\}$. The prime numbers between 10 and 40, are: $\{11, 13, 17, 19, 23, 29, 31, 37\}$. We choose the prime numbers which yield integer x and y.

13) Find the integer solutions of the equation: $4y = x^3 + 3x + 4$.

$$\left(\textbf{Ans:} \begin{cases} x = 4k \\ y = 16k^3 + 3k + 1 \end{cases} \textbf{or} \begin{cases} x = 4k - 1 \\ y = 16k^3 - 12k^2 + 6k \end{cases} \textbf{or}\right.$$

$$\begin{cases} x = 4k + 1 \\ y = 16k^3 + 12k^2 + 6k + 2 \end{cases}, \quad k \in \mathbb{Z}).$$

Hint: $y = 4q + r$, where $r = 0,1,2,3$. The given equation implies that $(x^3 + 3x + 4)$ is divisible by 4. We find for which $r \in (0,1,2,3)$ the number 4 divides exactly $(x^3 + 3x + 4)$.

14) Solve the Diophantine equation: $x^2 + 3(y^2 - 1) = 2025$.

(**Ans:** $(x, y) = (45,1), (0,26), (21,23), (24,22), (39,13)$).

Hint: $x^2 + 3(y^2 - 1) = 2025 = 3^2 \cdot 15^2$. The number 3 must therefore divide x, i.e. $x = 3k$, and the equation becomes: $3k^2 + (y^2 - 1) = 3 \cdot 15^2$. Similarly, it follows that 3 must divide $(y^2 - 1) = (y + 1)(y - 1)$, i.e. y must divide $(y + 1)$, or, y must divide $(y - 1)$. In the first case, $y = 3\lambda - 1$, in the second case, $y = 3\lambda + 1$, etc.

15) Find the right triangles, whose sides are integers, having the property: The area and the perimeter of the triangle are expressed by the same number.

(**Ans:** If x, y are the perpendicular sides and z is the hypotenuse:

$(x, y, z) = (12,5,13), (8,6,10), (6,8,10), (5,12,13)$).

Hint: We seek for positive integers x, y, z such that: $xy/2 = x + y + z$, i.e. $xy/2 = x + y + \sqrt{x^2 + y^2}$, or, $\sqrt{x^2 + y^2} = xy/2 - (x + y)$, square both sides, etc.

16) Solve the Diophantine equation: $x^2 = y^2 + z^2 - yz$.

(Ans: $x = (k^2 + k + 1)\lambda, \ y = (1 - k^2)\lambda, \ z = (1 + 2k)\lambda, \quad k, \ \lambda \in \mathbb{Z})$.

Hint: Set, $x = y + kz$.

17) Solve the Diophantine equation: $z^2 + 3w^2 + 4z + w = 21$.

(Ans: $(z, w) = (-1, -3), (-3, -3), (3, 0), (-7, 0))$.

Hint: The sought for numbers z and w will be the integer roots of the quadratic equation: $z^2 + 4z + 3w^2 + w - 21 = 0$. A necessary condition for this is the discriminant of the equation to be positive or zero **and** a perfect square. The discriminant $\Delta = 4^2 - 4(3w^2 + w - 21) \geq 0$, i.e. $3w^2 + w - 25 \leq 0$, which is satisfied for $-3.05 \leq w \leq 2.72$. The integer values of w, within this interval, which make Δ a perfect square, are: $w = -3$, or, $w = 0$, etc.

18) Solve the Diophantine equation: $x^2 - y^2 + 1 = 0$.

(Ans: $(x, y) = (0, \pm 1))$.

Hint: $(x + y)(x - y) = -1$, i.e. $(x + y) = 1, (x - y) = -1$, or, $(x + y) = -1, (x - y) = 1$.

19) Find the integers x, y, z which satisfy the equation:

$$xyz - xy + 2xz - 3yz = 2x - 3y + 6z - 6$$

(Ans: $(x = 3, y = k, z = \lambda), (x = k, y = -2, z = \lambda), (x = k, y = \lambda, z = 1))$.

Hint: The given equation is written as: $(x - 3)(y + 2)(z - 1) = 0$.

20) If $a, b \in \mathbb{Z}, k \in \mathbb{N}$, and $a^2 - kb^2 = 1$, show that: $1 < |a/b| \leq \sqrt{k + 1}$, (assume $ab \neq 0$).

Hint: $a^2 - kb^2 = 1$, i.e. $a^2 = 1 + kb^2 > kb^2$, i.e.$(a/b)^2 > k \geq 1$, i.e. $|a/b| > 1$. Also, from $a^2 = 1 + kb^2$, it follows, $a^2 \leq b^2 + kb^2$, i.e. $a^2 \leq (k+1)b^2$, i.e. $|a/b| \leq \sqrt{k+1}$.

21) Solve the Diophantine equation: $x^2 - (y-4)^2 = 11$.

(Ans: $(x, y) = (6,9), (6, -1), (-6, -1), (-6,9))$.

Hint: $(x + y - 4)(x - y + 4) = 11$, and since 11 is a prime number, $(x + y - 4) = 11 \ (or - 11)$ and $(x - y + 4) = 1 (or - 1)$, **or**, $(x + y - 4) = 1 \ (or - 1)$ and $(x - y + 4) = 17 (or - 17))$.

22) Find $c \in \mathbb{Z}$, so that the equation $11x + 7y = c$ to have exactly four, non negative integer solutions. Which are these solutions?

Hint: See Problem 5.

23) If $x^2 + 2y^2 = z^2$, show that $z = k^2 + 2\lambda^2$, $(x, y, z, k, \lambda \in \mathbb{Z})$. The numbers x, y, z are assumed to be relatively prime in pairs.

24) Solve the Diophantine equation: $x^2 + y^2 = z^2 + w^2$, $(x, y, z, w \in \mathbb{Z})$.

(Ans:

$$\left. \begin{cases} x = ac + bd \\ y = ad - bc \\ z = ac - bd \\ w = ad + bc \end{cases} \right\} \quad a, b, c, d \in \mathbb{Z}$$

Of course, since the equation is homogeneous, any multiple of x, y, z, w will also be a solution).

Hint: From the identity:

$$(ac \pm bd)^2 + (ad \mp bc)^2 = (a^2 + b^2)(c^2 + d^2)$$

If we set: $x = ac + bd, y = ad - bc, z = ac - bd, w = ad + bc$, the given equation is satisfied for any $a, b, c, d \in \mathbb{Z}$).

25) Solve the Diophantine equation: $7x^2 - 3x - 4y = 10$.

(Ans: $(x, y) = (4k + 2, 28k^2 + 25k + 3)$, or,
$(x, y) = (4k + 3, 28k^2 + 39k + 11))$.

Hint: From the given equation we have: $4y = 7x^2 - 3x - 10$. It follows that $7x^2 - 3x - 10$ must be a multiple of 4. We may write, $x = 4k + r$, where $r = 0,1,2,3$. We find that $4 / (7x^2 - 3x - 10)$, only when $r = 2$, or, $r = 3$, etc.

26) Find the integer solutions (x, y, z, w) of the system:

$$\{xy = z^2, \quad y + w = 21, \quad 2y = z + w\}$$

(Ans: $w \in A = \{22, 24, 28, 42, 84, 168, 30, 70, 462\}$, or, $w \in B = \{20, 18, 14, 0, -42, -126, 12, -28, -420\}$. The elements of the set A are the values of w which make the integer $(w - 21)$ to be a positive divisor of $441 = 21^2 = 3^2 \cdot 7^2$. The elements of the set B are the values of w which make $(w - 21)$ to be a negative divisor of 441. For a value of $w \in A \cup B$, the associated y, z and x are given by the formulas:

$$\left\{ \begin{array}{c} y = 21 - w \\ z = 42 - 3w \\ x = -9w + 63 - \dfrac{441}{w - 21} \end{array} \right\}$$

For example, for $w = 12$, we find: $y = 9, z = 6, x = 4$, for $w = 30$ we find, $y = -9, z = -48, x = -256$, etc.

Hint: From the second and the third equations of the given system we find: $y = 21 - w, z = 42 - 3w$, and then, from the first equation:

$$x = \frac{z^2}{y} = \frac{(42 - 3w)^2}{-w + 21} = \frac{9w^2 - 252w + 1764}{-w + 21} \Longrightarrow$$

$$x = -9w + 63 - \frac{441}{w - 21} \qquad (*)$$

Since x and w are integers, the number $(w - 21)$must be one of the divisors of $441 = 21^2 = 3^2 \cdot 7^2$. The divisors of 441 are: $\pm 1, \pm 3, \pm 7, \pm 3 \cdot 7$,

$\pm 3^2 \cdot 7,\ \pm 3 \cdot 7^2,\ \pm 3^2,\ \pm 7^2,\ \pm 3^2 \cdot 7^2$. Thus, for $w - 21 = 1$, we find $w = 22$, for $w - 21 = 3$, we find $w = 24$, etc.

27) Solve the Diophantine system:

$$\left\{ x + y = 2z, \quad xy = w^2, \quad \frac{1}{x} + \frac{1}{y} = \frac{5}{9} \right\}$$

(Ans: $(x, y, z, w) = (2,18,10,6), (18,2,10,6))$.

28) Solve the Diophantine equation: $z^2 + 6zw + 9w^2 = 26 + z - w$.

(Ans: $\{z = 12\lambda^2 + 13\lambda - 16,\ w = -4\lambda^2 - 3\lambda + 6\}$, or $\{z = 12\lambda^2 + 19\lambda - 12,\ w = -4\lambda^2 - 5\lambda + 5\}$, with $\lambda \in \mathbb{Z}$).

Hint: $(z + 3w)^2 - (z - w) = 26$, or if we set $x = z + 3w, y = z - w$, $x^2 - y = 26$, the solution of which is $\{x = k, y = k^2 - 6,\ k \in \mathbb{Z}\}$, and going back to the variables z and w, we find: $\{z + 3w = k,\ z - w = k^2 - 6\}$. Solving for w we find: $4w = k - k^2 + 26$. Since w is an integer, $k - k^2 + 26$ must be a multiple of 4. In general, $k = 4\lambda + r$, where $r = 0,1,2,3$. Substituting this expression of k in the expression of w, we find:

$$4w = \underbrace{-16\lambda^2 + 4\lambda(1 - 2r)}_{mul.4} + r - r^2 + 26 = mul.4 + \underbrace{r - r^2 + 26}_{a}$$

We find that $4w$ is a multiple of multiple of 4, provided that $a = r - r^2 + 26$ is a multiple of 4, and this occurs when $r = 2$ and $r = 3$, etc.

29) Find the integers x and y that satisfy the equation:

$$z(3 - |w|) + w(3 - |z|) + |zw| = 6$$

(Ans: $(z, w) = (4,6), (6,4), (2,0), (0,2), (-2,-4), (-4,-2))$.

30) Find the integer solutions x, y of the equation:

$$y^{2x} - x^{2y} = 17, \quad x, y > 0$$

(Ans: $(x, y) = (2,3))$.

Hint: $y^{2x} - x^{2y} = (y^x)^2 - (x^y)^2 = (y^x - x^y)(y^x + x^y) = 17$, and since 17 is a prime number, we must have: $(y^x - x^y) = 1, (y^x + x^y) = 17$, or, $(y^x - x^y) = 17, (y^x + x^y) = 1$, etc.

31) Find the integer and positive solutions of the equation:

$$x^3 + y^3 = (x + y)(10x + y)$$

(Ans: $(x, y) = (3,7), (4,8), (10,11), (11,11), (11,1), (14,8), (14,7))$.

Hint: Since $x + y > 0$, the given equation reduces to: $x^2 - xy + y^2 = 10x + y$.

32) Find the integer solutions of the equation: $3(x^2 + y^2) + 2xy = 132$.

(Ans:
$(x, y) = (7, -3), (3, -7), (-3,7), (-7,3), (5,3), (-3, -5), (3,5), (-5, -3))$.

Hint: The given equation is written as: $2(x + y)^2 + (x - y)^2 = 132$, or, if we set, $X = x + y$, $Y = x - y$, $2X^2 + Y^2 = 132$. The integer values of X and Y which satisfy this equation are: $(|X| = 4, |Y| = 10)$, or, $(|X| = 8, |Y| = 2)$, and from these values of X and Y, we find x and y.

33) Find the integer numbers y, z, w which satisfy the relations:

$$\left\{ \begin{array}{l} 35y + 63z + 45w = 1 \\ |y| < 9, |z| < 5, |w| < 7 \end{array} \right\}$$

(Ans: $(y, z, w) = (8, -3, -2), (-1, -3,5))$.

Hint: The general solution is:

$$\left\{ \begin{array}{l} y = 17 - 27k - 9\lambda \\ z = -3 + 5k \\ w = -9 + 14k + 7\lambda \end{array} \right\} \quad k, \lambda \in \mathbb{Z}$$

34) If k, λ are relatively prime integers and $(x_1, y_1), (x_2, y_2)$ are two integer solutions of the equation $kx + \lambda y = c, (c \in \mathbb{Z})$, then, if $|y_1 - y_2| < |k|$, show that: $x_1 = x_2$ and $y_1 = y_2$.

Hint: From $kx_1 + \lambda y_1 = c$ and $kx_2 + \lambda y_2 = c$, it follows that: $k(x_1 - x_2) = -\lambda(y_1 - y_2)$, i.e. $|k||x_1 - x_2| = |\lambda||y_1 - y_2|$. If $(y_1 - y_2) \neq 0$, then, since k and λ are relatively prime, $|k|$ must divide $|y_1 - y_2|$, which cannot be true since $|y_1 - y_2| < k$. Thus, necessarily, $(y_1 - y_2)$ must be zero, etc.

35) Show that the Diophantine equation $ax + by = c$, $(a, b, c \in \mathbb{Z})$ does not have positive solutions when $ab > 0$ and $ac \leq 0$.

Hint: If the equation had a positive solution (x_0, y_0), then: $ax_0 + by_0 = c$, and multiplying both sides by a, we would have, $a^2 x_0 + aby_0 = ac$, which cannot be true, since $a^2 x_0 + aby_0 > 0$ while $ac \leq 0$.

36) Solve the Diophantine equation: $(a + 3)x - (a + 2)y = 2$, $(a \in \mathbb{Z})$.

(Ans: If $a = -2$, the solution is: $\{x = 2, y = k \ (arbitrary \ ineger)\}$, if $a = -3$, the solution is $\{x = \lambda \ (arbitrary \ ineger), y = 2\}$. If $(a + 2)(a + 3) \neq 0$, the solution is: $\{x = 2 + (a + 2)m, y = 2 + (a + 3)m\}$, where $m \in \mathbb{Z}$.

37) Show that the equation $3y^2 + 11 = x^2$ does not have integer solutions.

Hint: $x = 3k + r$, where $k \in \mathbb{Z}$ and $r \in \{0,1,2\}$.

38) Find the positive integers x, y which satisfy the equation: $3^x - 2^y = 1$.

(Ans: $(x, y) = (2,3), (1,1)$).

Hint: Consider the two cases: **a)** $x = 2k$ (even), and **b)** $x = 2k + 1$, (odd).

39) Find the solutions of the equation $x^2 + y^2 = z^2$, which satisfy the relation: $z = y + 1$.

(Ans: $x = 2k + 1$, $y = 2k^2 + 2k$, $z = 2k^2 + 2k + 1$, or $x = -2k + 1$, $y = 2k^2 - 2k$, $z = 2k^2 - 2k + 1$).

40) a) Find the positive and integer solutions (a, b, c) of the Pythagorean equation $x^2 + y^2 = z^2$ which form an arithmetic progression, **b)** Are there any positive and integer solutions of the same equation which form a geometric progression?

(**Ans: a)** $a = 3n^2, b = 4n^2, c = 5n^2, n = 1,2,3, ...$, **b)** There are no such solutions).

Hint: a) a, b, c form an arithmetic progression if $2b = a + c$, **b)** a, b, c form a geometric progression if $b^2 = ac$, with $a = m^2 - n^2, b = 2mn, c = m^2 + n^2$ (the solutions of the Pythagorean equation). The equation $b^2 = ac$ implies that

$$4m^2n^2 = m^4 - n^4 \Leftrightarrow m^4 - n^4 - 4m^2n^2 = 0 \qquad (*)$$

This is a homogeneous equation, (all its terms ore of the same degree 4), and if we divide both sides by n^4, we find:

$$\left(\frac{m}{n}\right)^4 - 4\left(\frac{m}{n}\right)^2 - 1 = 0 \overset{\left(t=\left(\frac{m}{n}\right)^2\right)}{\Longleftrightarrow} t^2 - 4t - 1 = 0 \qquad (**)$$

For m and n integers, the number $t = (m/n)^2$ must be a rational number. However, solving eq. (**) for t we find:

$$t = \frac{-(-4) \pm \sqrt{(-4)^2 - 4 \cdot 1 \cdot (-1)}}{2} = \frac{4 \pm \sqrt{20}}{2} = 2 \pm \sqrt{5}$$

Due to the number $\sqrt{5}$, t is **an irrational number**, i.e. there are no integer numbers m and n to satisfy eq. (**), and this means that the solutions of a Pythagorean equation cannot form a geometric progression.

www.ingramcontent.com/pod-product-compliance
Lightning Source LLC
Chambersburg PA
CBHW080958290526
45795CB00009B/2994